QUANTITATIVE
PLANT ECOLOGY

ENGLAND:	BUTTERWORTH & CO. (PUBLISHERS) LTD.
	LONDON: 88 Kingsway, W.C.2
AUSTRALIA:	BUTTERWORTH & CO. (AUSTRALIA) LTD.
	SYDNEY: 20 Loftus Street
	MELBOURNE: 473 Bourke Street
	BRISBANE: 240 Queen Street
CANADA:	BUTTERWORTH & CO. (CANADA) LTD.
	TORONTO: 1367 Danforth Avenue, 6
NEW ZEALAND:	BUTTERWORTH & CO. (NEW ZEALAND) LTD.
	WELLINGTON: 49/51 Ballance Street
	AUCKLAND: 35 High Street
SOUTH AFRICA:	BUTTERWORTH & CO. (SOUTH AFRICA) LTD.
	DURBAN: 33/35 Beach Grove
U.S.A.:	BUTTERWORTH INC.
	WASHINGTON, D.C. 20014: 7300 Pearl Street

QUANTITATIVE PLANT ECOLOGY

P. Greig-Smith, M.A. (Cantab.)

Reader in Botany, University College of North Wales

SECOND EDITION

LONDON
BUTTERWORTHS
1964

Suggested U.D.C. No. 581.5.001.5

First published 1957
Second edition   1964
Reprinted 1965

Printed and bound in Great Britain by
The Garden City Press Limited
Letchworth, Hertfordshire

CONTENTS

PREFACE TO SECOND EDITION

In the six years since the first edition was written there has been a steady interest in quantitative methods in plant ecology. An encouraging feature has been the appearance of an increasing number of publications in which the emphasis has been on the use of quantitative methods rather than on the methods themselves. At the same time there have been considerable advances in methodology, which have necessitated substantial revision for this edition. Development of methods has been particularly active in classification and ordination of communities. The account of this topic has been very largely rewritten and now appears as a separate chapter (Chapter 7). Substantial additions have been made to the remaining chapters. One of the appendix tables has been expanded and two further tables included. It has been necessary, on account of the increasing volume of literature, to be more selective in citing references than in the first edition, but I believe the reference list again includes the majority of references, up to the end of 1962, which are likely to be useful to anyone concerned with choice of methods. Goodall (1962) has recently provided a more comprehensive bibliography, including papers reporting results based on the use of quantitative methods.

I am indebted to Professor W. T. Williams and Dr. J. M. Lambert for making their account of nodal analysis available before publication and for reading a draft of Chapter 7 and making various helpful suggestions. I am particularly grateful to Professor Williams for allowing me to quote from work still in progress by himself and Mr. M. B. Dale. I am also grateful to Dr. D. J. Anderson, Dr. A. W. Ghent and Mr. R. T. Gittins for allowing me to quote conclusions in advance of their publication, to Mr. R. I. J. Tully and his assistants, who have again been most helpful in tracing references, and to Mrs. P. M. Venesoen, who has typed successive drafts. My wife has again given me invaluable assistance in the preparation of the manuscript for the press.

Bangor   P. GREIG-SMITH
    May, 1963

PREFACE TO FIRST EDITION

CHANGE from a qualitative to a quantitative approach is characteristic of the development of any branch of science. As some understanding is achieved of the broader aspects of phenomena, interest naturally turns to the finer detail of structure or behaviour, in which the observable differences are smaller and can only be appreciated in terms of measurement. It is not surprising that a quantitative outlook has been attained earlier in most branches of physical science than in biological science. Perhaps the greatest single difference in methodology between the physical and biological sciences is that in the former it is generally possible to isolate one variable at a time for study, whereas in the latter this is rarely possible. Thus, in the physical sciences broad outlines of phenomena are more readily seen from a relatively simple programme of qualitative investigation, and the way cleared for the more exact quantitative approach. In biology not only is it rarely possible to isolate variables for study, but the subjects of investigation are themselves commonly so complex that they are difficult to measure. In some branches of biology, therefore, attainment of the quantitative stage is perhaps little more than an ideal unlikely to be achieved in the near future. Other branches are more tractable, notably plant physiology, where many 'unwanted' variables can at least be minimized by use of controlled environments and clonal material, and investigation is now very largely in the quantitative stage. Plant ecology is at present in a transitional stage, and great advances can be expected from the quantitative techniques now being developed.

The general impossibility of controlling 'unwanted' variables in biology leads to a much greater degree of error variability in measurement than in the physical sciences. In the physical sciences differences among replicate measurements are generally attributable to deficiencies of technique, whereas in biological observations differences may be due not only to these deficiencies, but also, and commonly to a much greater extent, to fluctuation in variables not under investigation and assumed to be constant. Put another way, it is very much more difficult to obtain truly replicate samples in biological measurements than in physical measurements. If measurements are made in two or more contexts with the object of determining if there is any difference in the variable measured, the means may be different but the ranges of individual measurements may overlap. Thus the problem arises whether an observed difference is significant or not, i.e. whether it reflects any real difference between the two groups sampled, or is due to chance. In the physical sciences the immediate reaction is to improve the technique to obtain more accurate measurements, when either the means remain different but the ranges no longer overlap, indicating real difference, or the means converge with ranges still overlapping, suggesting that there is no real difference. In biology, however, there is often little scope for improvement of technique, and the biologist is therefore forced to turn for help in judging significance of difference to the techniques of statistical analysis. These are based on probability theory and permit determination of the probability of observed differences arising by chance in different samples of the same population. Thus arises the apparent paradox that while the physical sciences make much greater use of a quantitative approach than the biological sciences, they are much less dependent on the techniques of statistical analysis for the interpretation of their quantitative data.

There is at the present time a growing awareness among ecologists of the need to place their science on a more exact basis. The impetus given by the pioneers of ecology, which led to rapid advances in the first three or four decades of this century, is dying down and it is clear that the emphasis is changing from extensive work on vegetation to intensive work on selected aspects. If this is to go forward, more exact techniques of examination are necessary. The need for improved technique is made the more urgent on the one hand by the rapid depletion of natural vegetation, the only source of data on many of the more fundamental aspects of ecology, and on the other hand by the realization that advances in many branches of agriculture and forestry depend upon answers to ecological questions.

Although aware of the value of the quantitative approach and of the valuable tools available in the techniques of statistical analysis, many ecologists, faced with the rapidly expanding literature on quantitative methods in ecology, are reluctant to adopt a fully quantitative approach. This reluctance is, perhaps, reinforced by the apparent disregard of practical ecological problems in some theoretical studies. There is a need, therefore, for an assessment of the practical potentialities of various methods and techniques. I have attempted to make such an assessment in this book, and it is hoped that an ecologist faced with a particular problem in the field will find here guidance on the most profitable means of obtaining and handling his data, as well as a broad survey of the quantitative approach to plant ecology.

Chapter 1 is concerned with the different methods of describing vegetation in quantitative terms. Chapter 2 follows closely on the first, and deals with the positioning and number of samples to be used and with the comparison of the results of different sets of observations.

In Chapter 3 a hypothesis is developed of the significance, in relation to determining factors, of pattern, i.e. departure from randomness of distribution of individuals within the plant community. This leads to consideration of the techniques of detection and analysis of pattern. Chapter 4, on association between species, considers pattern from another aspect, the relationship between the patterns of different species.

Chapter 5 deals with correlation between vegetation and the level of environmental factors, the type of data on which most conclusions on the main factors determining the distribution of plants have been based.

Chapter 6 is concerned with the delineation of plant communities and assessment of difference between vegetation stands. This inevitably leads to consideration of the classification of vegetation, whether it is possible and if so, how it may be placed on an objective basis.

Chapter 7 is to some extent speculative. Believing that the quantitative approach has its own distinctive contribution to make to ecological theory, as well as putting existing practices on a sounder basis, I have devoted the greater part of this final chapter to consideration of this theme.

The appendices include brief discussions of the handling of meteorological data and of the area occupied by species, topics which do not fit conveniently into the main text, and a few tables of functions which are not readily available.

I have assumed that the reader, if he intends to make serious use of quantitative methods, will have access to Fisher and Yates' 'Statistical Tables for Biological, Agricultural and Medical Research' or to some other source of the commoner statistical tables, and to one or other of the elementary textbooks of statistical methods for biologists such as Mather's 'Statistical Analysis in Biology' or Snedecor's 'Statistical Methods Applied to Experiments in

Agriculture and Biology'. At the same time I have not hesitated to illustrate and discuss procedures at length where experience suggests that the biologist who is not very mathematically minded has difficulty in grasping them. By contrast I have given no examples of computation at all for more complicated statistical procedures, but have merely outlined their principles, believing that in such cases the biologist should take advice from a competent statistician, at least until he has had very considerable experience of statistical methods.

The reference list is strictly a list of references cited and makes no attempt at completeness. At the same time I believe that it includes the majority of references on methodology which are of current practical value rather than historical interest. The literature pertaining to the subject matter of Chapters 1 to 4 and Chapter 6 has been listed fairly completely by Goodall (1952a).

The quantitative ecologist is very dependent upon sound statistical advice. It therefore gives me great pleasure to express here my especial thanks to Professor M. S. Bartlett, of the University of Manchester, who has given freely of his time to advise me on statistical methods on various occasions since I first became interested in quantitative ecology. He has very generously read the whole of this book in manuscript, except Chapter 7, and corrected a number of mis-statements and ambiguities.

To Professor P. W. Richards I am indebted for his constant encouragement during the writing of the book and for reading parts in manuscript and making a number of suggestions for improvement. I am grateful to Mr. R. I. J. Tully, of the Library of the University College of North Wales, for help in obtaining literature from other libraries, and to Dr. A. D. Q. Agnew, Dr. K. A. Kershaw and Dr. W. S. Lacey for allowing me to quote data prior to their publication.

Lastly, I am grateful to my wife for much help in preparation of the manuscript for publication.

P. GREIG-SMITH

Bangor
April, 1957

QUANTITATIVE DESCRIPTION OF VEGETATION

MANY ecological data take the form of description of vegetation with or without concurrent recording of factors of the environment. Such data have formed the main basis of most ecological theory and concepts and are likely to continue to do so. It is essential, therefore, to place the description of vegetation on as sound a basis as possible. Two methods of community description have, with only minor modifications, predominated in the past. The first involves the making of a complete list of the species present in a community with the assignment of 'frequency symbols' or numerical ratings by inspection. This developed from the subjective assessment of species as rare, occasional, common, etc. in floras and represents an essentially similar process applied to a much smaller and more closely defined area. The second method derives from the work of Raunkiaer (1909, see Raunkiaer, 1934), and depends on the recording of presence or absence of species in small samples of the community under investigation. The sampling unit is a square, or less commonly a rectangle or a circle, of defined area, which may be placed either at random or in some regular manner. The results are expressed as the *percentage frequency*, i.e. the percentage of samples in which each species has been found. The species may be grouped for convenience of comparison into a number of frequency classes, with the limits of the classes forming either an arithmetic or a geometric series. The five classes 0–20 per cent, 20–40 per cent, 40–60 per cent, 60–80 per cent, and 80–100 per cent have commonly been used. This procedure is sometimes referred to as *valence analysis*.

These two methods have been widely used and it is important to consider their validity and limitations. The species list with frequency symbols or ratings is so well established that its value is too rarely questioned. The surprising feature is the degree of consistency of results obtained by experienced field workers. Several attempts have been made to assess the importance of the personal factor in deciding the rating assigned. Hope-Simpson (1940) has shown that one observer may give markedly different

1

assessments on different occasions, particularly at different seasons. Smith (1944) investigated personal error affecting the simpler technique of visual assessment of percentage cover of total vegetation in plots. He showed that individual observers out of a group of eight deviated in their assessments from the group mean by as much as 25 per cent. There was little evidence of any tendency to give consistently lower or higher values than the mean. There is little doubt that similar results would be obtained if the assignments of frequency symbols by a number of observers for the same sample area were compared. Indeed, in view of the complex of factors affecting assignment of a symbol, considered below, the discrepancy might well be greater. A further source of personal error lies in the mental state of the observer. Every ecologist with some experience of frequency estimation is aware that rare and inconspicuous species tend to be rated lower when the observer is tired than when he is fresh and fully alert. Conversely, familiarity with a vegetation type and the species involved tends to produce higher ratings.

The difficulties introduced by personal factors are perhaps less important to an independent worker than to teams of workers. Once he is sufficiently experienced to attain reasonable consistency in repeated assessments of the same vegetation his ratings are likely to give a fairly reliable comparison between different communities. It is still impossible to attach any absolute value to his results. Moreover, results of different workers cannot be compared except in very broad terms, unless ratings for at least one and preferably several communities in common are available for the several observers. This drawback is so serious that, in general, frequency symbols should be used as the sole description of a community only when lack of time prevents the use of any more exact measure.

The difficulties of comparison of the results of several workers are more obvious in co-operative work. Unfortunately it is often in such work, e.g. in broad-scale surveys of the vegetation of large areas, that a rapid method of description is required. The errors introduced by personal factors can, however, be considerably reduced by careful standardization on the same vegetation between members of a team before the work is started and at intervals during its course.

There is a more serious objection to the use of frequency symbols. Several factors influence the observer in his assignment of a frequency symbol. Those uppermost in the minds of most observers are probably *density* or number of plant units per unit area and *cover* or percentage of the total area covered by the aerial parts of plants of a species, rather than true *frequency*, which is itself a complex character (see below). It is, however, difficult to avoid being influenced by the differing growth forms of different species and by the varying pattern of distribution of the individuals of different species on the ground, two factors greatly affecting the relative conspicuousness of different species. Even if density and cover alone are taken into consideration, an ideal probably impossible to attain, an attempt is being made to assess on one scale two largely independent variables. The problem is made clear by inspecting a weedy lawn, which includes at one extreme grasses, such as *Poa annua*, which have a high density of shoots but relatively low cover value per shoot, and at the other rosette weeds such as *Plantago* spp., each shoot of which covers a relatively large amount of ground. Cover and density each represent only one aspect among several of the contribution made by a species to the community. With practice in one vegetation type an observer can establish for himself an arbitrary scale of relative importance attached to cover and density but this relationship can scarcely be standardized and cannot readily be communicated to others. Moreover it has to be established afresh for communities of different physiognomy.

The confusion between cover and density is unavoidable. Other sources of error may with experience be reduced, though not eliminated. Species vary in conspicuousness and it is difficult to avoid overrating conspicuous species and underrating inconspicuous ones. Even the same species may vary greatly in conspicuousness between flowering and non-flowering states, e.g. *Deschampsia flexuosa* growing in comparatively small quantity amongst *Eriophorum vaginatum* gives a distinctive appearance to the community when flowering but is picked out only with difficulty when purely vegetative. Seasonal variation in assessment by the same observer on the same community, other than for annual species, results mainly from this variation in conspicuousness between different states. A further complication is introduced by the varying patterns that individuals of different species may form within the community. Individuals may be distributed more or less randomly through the community or they may be markedly aggregated into groups more or less clearly separated by areas in which the species is lacking or sparse. A high degree of aggregation may be indicated by rating the species as 'locally abundant', 'locally frequent', etc. When the aggregation is less marked overrating is liable to occur through increased conspicuousness of a species whose individuals are found in groups. The nature of these spatial patterns is of great importance to an understanding of community structure and is discussed in detail in a later chapter.

3

The use of frequency symbols has been treated at some length because their nature is often not fully understood, and because the method is commonly regarded as an elementary and straightforward one not needing any detailed consideration. It is evident, however, once its basis is examined, that care and experience are necessary before results of value can be obtained and that at its best the method is subject to considerable error. It is more satisfactory in description and comparison of communities of similar and uniform physiognomy than in those including diverse growth forms.

Frequency symbols raise a further point of importance: the highest category normally used is 'dominant'. This term is rather an unfortunate one, but probably too well established to be replaced. In practice it generally represents nothing more than the highest grade of density plus cover in the vegetation under examination. Many ecologists accept to a greater or lesser extent an organismal view of the plant community and tend to confuse this use of the term 'dominant' with the concept of dominance as a degree of influence exerted over other species of the community (by a variety of competition and stimulant effects). It is clear that some degree of such controlling influence has often been attached to species figuring as 'dominant' in species lists, without any evidence other than their having been assigned the highest frequency rating. A species may be dominant in both senses in a community but is not necessarily so. Moreover, the existence of dominance in the second sense is not universally accepted. It would certainly lead to clarity if another term could be substituted for 'dominant' as a frequency symbol. The dominant species of a community might then be defined as that species which exerts the greatest influence on other species of the community and is least influenced by them. The determination of the dominant species of a community in this restricted sense would involve prolonged investigations in most, if not all, communities, including autecological studies of all the more important species. It would probably be least difficult in forest with a single species only in the canopy layer. A third sense for 'dominant' is occasionally found in ecological literature, viz. to describe any individual tree of the canopy in forest whose crown is more than half exposed to full illumination. It is widely used in this sense by foresters together with the complementary term 'predominant' (better, 'emergent') for individuals rising above the continuous canopy. Richards *et al.* (1940) suggested that 'dominant (ecol.)' should be used for the species with the highest frequency rating where there is any doubt about the meaning intended.

The percentage frequency method is more conveniently considered after discussion of the various quantitative measures of vegetation. There are many such measures and it will be necessary to consider in detail only the more important. They fall into two categories, those in which the figure obtained is independent, within the limits of observational error, of the method used to determine it, and those in which it is dependent on the mode of sampling and has meaning only when coupled with a statement of the method used. Of the former, which may be described as *absolute measures*, the more important are density, cover and the various measures of yield. Of the *non-absolute measures* the only one of importance is frequency.

Density is the measure of number per unit area. The objects enumerated may be either whole plants or portions of a plant, depending on the morphology of the species involved. Thus individuals of trees or annual herbs are usually clearly distinguishable but definition of an individual in many perennial herbs is difficult or impossible, e.g. many rhizomatous and rosette-forming species. Even if individuals are recognizable they may not be the most useful unit owing to their wide range of size, e.g. tussocking grasses in which tillers are more appropriate units. The term *mean area*, introduced by Kylin (1926) and defined as the reciprocal of density, is sometimes useful. Density is readily determined by direct counts in suitable sample areas. Density is inversely related to the mean distance between individuals and attention has recently been paid to the possibility of using a measure of this distance to estimate density, to avoid the necessity of laying out sample areas (see Cottam and Curtis, 1956). Unfortunately the determination of the mean distance is complicated by the spatial arrangement of individuals. The difficulties involved will be discussed in Chapter 2.

Cover is defined as the proportion of ground occupied by perpendicular projection on to it of the aerial parts of individuals of the species under consideration. Its nature is perhaps most clearly brought out by noting that if a community on level ground composed of one species only were illuminated vertically the proportion of ground in shadow would represent the cover of the species. Cover is usually expressed as a percentage, and it should be noted that the total cover for all species in a community may exceed 100 per cent and normally does so in all except open communities. This follows from the overlying or underlying of a part of one individual by parts of one or more others of different species, obvious in any closed community.

Cover may be either estimated or measured. Estimations are subject to the personal errors already discussed in relation to frequency symbols, though easier to make since they involve a

5

single characteristic only. Various techniques have been devised to assist in estimation of cover. The literature has recently been reviewed by Brown (1954). Measurement of cover may be made by the point quadrat method, which depends on recording the presence or absence of a species vertically above a number of points in the community being described. The percentage of points above which the species is present represents the percentage cover. The theoretical basis of this method is simple. If a sample area has a finite but small size it may be completely covered by the projection of the aerial parts of individuals of a species under consideration, incompletely covered or not covered at all. As the size of the sample area is reduced it becomes more likely that it is either completely covered or not covered at all, until, when it is infinitely small, i.e. a point, it is always either completely covered or not covered. As long as the sample area is finite the whole area under examination may be considered as consisting of a large but finite number of such sample areas, each falling into one of the categories completely covered, partially covered and not covered. As the sample area decreases in size the proportion of the total number of sample areas which are either partially or completely covered approaches more nearly to the value of the species cover. At the limit, when the sample area becomes a point, the proportion which is covered of the infinitely large total number of sample areas equals the cover of the species. The points actually examined in an investigation, if properly selected, are an unbiased sample of the infinitely large number of possible points and give an estimate of the true value of cover, the accuracy of which can be increased to any desired degree of precision by increasing the sample size.

In practice the sample size used cannot, of course, be a true point. Warren Wilson (1959a, b) has used a fine needle mounted on the end of a rod, recording only contacts with the tip, and Winkworth and Goodall (1962) have described a sighting device incorporating two sets of cross wires; both methods give a very close approximation to a point. Sampling has, however, usually been by means of long pins which are lowered through the vegetation. The use of a pin of finite diameter will give a value of cover greater than the true value because plants will be touched that would not make contact with the axis of the pin. The magnitude of this effect is frequently not realized. It is demonstrated by the data in TABLE 1 and by FIGURE 1, both taken from Goodall (1952b), who gives a critical discussion of the point-quadrat method. The values given in the table for pin diameter 0 were obtained by an apparatus of cross wires.

Basal area, as generally understood, is a measure somewhat similar to cover, being the proportion of ground surface occupied by a species. (The term has also been used instead of cover, e.g. West, 1937.) It is of particular value in dealing with species of tussock form. Its estimation and measurement involve similar considerations to those of cover. In measurement presence of basal parts of the plant at the sampling point is substituted for presence of aerial parts.

An alternative method of measuring cover is to record total length of interception made by plants of a species on line transects.

TABLE 1

FREQUENCY (PER CENT) OF CONTACT BETWEEN FOLIAGE AND PINS OF DIFFERENT DIAMETERS (from Goodall, 1952b, by courtesy of *Aust. J. Sci. Res.*)

Locality	Species	No. of points	Pin diameter (mm)			Significance (P) for difference in pin diameter	
			0	1·84	4·75	0–1·84 mm	1·84–4·75 mm
Seaford	*Ammophila arenaria*	200	39·0	66·5	71·0	< 0·001	> 0·05
	Ammophila arenaria	200	60·5	74·0	82·0	0·001–0·01	> 0·05
Black Rock	*Ehrharta erecta*	200	74·5	87·0	93·5	0·001–0·01	0·01–0·05
Sorrento	*Lepidosperma concavum*	200	19·5	22·0	27·5	> 0·05	> 0·05
	Spinifex hirsutus	200	35·0	48·5	61·0	0·001–0·01	0·01–0·05
Carlton	*Fumaria officinalis*	200	20·5	31·5	30·0	0·01–0·05	> 0·05
	Ehrharta longiflora		24·5	25·5	37·5	> 0·05	0·01–0·05
	No contact		53·0	42·5	38·5	0·01–0·05	> 0·05
	Lolium perenne	200	65·0	85·5	82·5	< 0·001	> 0·05

The proportion of the total length of the transect intercepted by a species gives a measure of the cover of that species. This method has an evident advantage in speed of working and in eliminating the exaggeration introduced by using a point of finite size, but involves some approximation in that an individual plant must be assumed to have a definite boundary within which it has 100 per cent cover. It is thus well adapted to measurement of basal area, and of cover of densely tussocking species, where this condition is nearly true, but of little use for vegetation where species are intermingled with one another. It is also appropriate to the measurement of the canopy of trees and shrubs, a feature sometimes of importance and not necessarily equivalent to their cover. How nearly cover and canopy correspond depends on the proportion of gaps in the leaf mosaic.

Measures of *yield* call for little specific comment. They are determinations of quantity of material produced per unit area and are made in a similar way to density determinations, the harvest of the relevant material being substituted for a count of plant units. An

1cm

☐ 0 mm
▨ 1 mm
▬ 2 mm
▦ 3 mm
▮ 4 mm

FIGURE 1. Projection of part of tussock of *Ammophila arenaria*, showing areas over which contact would be made with the foliage by pins of the diameters stated; each zone is understood as including those with less dense shading (from Goodall, 1952b, by courtesy of *Aust. J. Sci. Res.*)

indirect estimate of bulk of shoot material can be readily obtained in some types of grassland by a modification of the point-quadrat method. If the total number of hits made on each species is recorded instead of only their presence or absence at each sampling point, it has been found empirically that the proportions of hits on the different species correspond closely to the weight of shoot material. The validity of interpreting the proportions of hits in this way

8

clearly depends on the species having similar growth form. Where it can be applied it is particularly valuable because it permits of an estimate of yield without destroying the vegetation, so that observations can be repeated on subsequent occasions without allowance having to be made for interference. Yield determinations have, naturally, been widely used in applied botany, e.g. dry weight in pasture assessment, volume of timber in forestry. Relatively much less use has been made of them in ecology, though they are valuable in such problems as the study of humus or mineral nutrient turnover, where it is important to know the rate of addition of plant material to the soil.

The measures so far considered are straightforward in conception, though they may present difficulties in determination. *Frequency* is usually the easiest of the quantitative measures to determine, but its meaning in biological terms is not so clear-cut. Frequency of a species, determined by a particular size of sample area, is the chance of finding the species within the sample area in any one trial. It is determined by examining a series of sample areas placed at random within the vegetation being described and recording the species present in each sample area. The number of samples in which a species occurs, expressed as a proportion or percentage of the total, is an estimate of the chance of its occurring in any one sample, i.e. the frequency. The accuracy of the estimate can be increased to any desired extent by increasing the number of samples. It is evident from the definition that a frequency value has meaning only in relation to the particular size and shape of sampling area used. Increase in size of sampling area will necessarily result in an increase in the chance of a species occurring in any particular sample.

The ease of determination of frequency, in comparison with density and cover, has weighed heavily with ecologists. Frequency has one disadvantage in that a value for one particular position on a transect line or a grid cannot be obtained by the normal method of random throws. A small quadrat in which a count of density is made, or a frame of, say, ten or twenty pins for cover determination may be placed in a definite position on a line or grid. The resulting value, though subject to a relatively large sample error, can be localized in relation to the frame of reference provided by the line or grid. Areas within which frequency is determined by random throws must be relatively larger, and the value obtained, though subject to proportionately less sample error than single determinations of density or cover, cannot be so precisely localized. Correlation with habitat factors which vary over small areas is thus made more difficult to detect. The difficulty may be largely overcome by using a closely similar measure which may be termed *local frequency*. Instead of a simple quadrat in which presence or absence is recorded, the sampling unit consists of a quadrat divided into a number of smaller squares, e.g. a 25 cm square quadrat divided into twenty-five squares each 5 cm × 5 cm. Presence or absence is recorded for each sub-unit. The value obtained, though subject to a relatively large sampling error, can be localized to an area considerably smaller than any within which random placing of quadrats would be practicable.

'Presence' of species in sample areas may be interpreted in two different ways. It is commonly taken to mean that an individual or part of an individual is rooted in the sample area. Alternatively the occurrence of any aerial part of the plant may be taken as indication of presence. When it is necessary to distinguish the two different usages they may be referred to as 'rooted frequency' and 'shoot frequency' respectively. Most species will clearly show a much higher shoot frequency than rooted frequency in the same community. In the following pages frequency will be taken to mean rooted frequency unless shoot frequency is specifically stated.

Rooted frequency is clearly not independent of the absolute measure of density. In two communities, otherwise comparable, one of which has a higher density of a particular species than the other, the frequency of the species will be higher in the community with the greater density. There is a similar relationship between shoot frequency and cover. Various attempts have been made to establish an exact relationship between frequency and density, both empirically and theoretically. Difficulty arises because frequency is dependent not only on the number of individuals in the area but also on the way in which they are distributed over the area or, in other words, on the pattern. The effect of pattern is illustrated diagrammatically in FIGURE 2. Each square contains the same number of dots, representing individual plants, so that density is the same in all three squares. In FIGURE 2(a) the individuals are uniformly spaced. If the community were sampled by a quadrat of the size shown, 100 per cent frequency would be found because the maximum distance between adjacent individuals is less than the shortest dimension of the quadrat. In (b) the opposite extreme is shown with all the individuals occurring close together in one part of the area; here sampling with the same size quadrat would give a low frequency value. FIGURE 2(c) represents an intermediate condition, more comparable to the type of situation commonly found in the field, with individuals forming a number of groups more or less clearly separated from one another; here sampling would give a frequency intermediate between the first two figures.

10

Frequency is thus dependent partly on density and partly on pattern. This has both advantages and disadvantages. There is the disadvantage found in the use of frequency symbols that two different properties are being assessed on the same scale, with the result that the same frequency may be shown by a species in different communities in which it plays quite different parts in the make-up of the vegetation. On the other hand frequency does integrate to some extent two important aspects of vegetation without any subjective comparison between them being necessary. If both aspects were readily assessed without much additional work being required, there would be little excuse for using frequency.

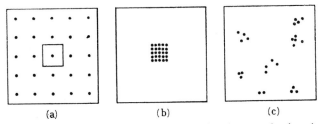

(a) (b) (c)

FIGURE 2. Three different distributions having the same density. A quadrat in position in (a) (see text)

Density, as we have already seen, is readily determined provided the species under consideration present units that can be counted, but the determination is often time-consuming. Pattern is never so regular that it can be completely described in any simple mathematical terms. Still, a close approximation can often be found, capable of expression in a relatively simple manner. Techniques of determination of pattern (to be considered in Chapter 3) are developing rapidly, but it seems likely that they will always be laborious compared with a simple frequency estimate. For many purposes the loss of possible information when frequency is used is counterbalanced by the great gain in speed of description of vegetation. Hence frequency determination is likely to remain an important ecological tool.

If the pattern of individuals is of a form that can be completely described in mathematical terms, then the corresponding frequency for a particular density and size of quadrat can be calculated, though it may be a laborious operation to do so, and, moreover, the result is of little practical value. If, however, the individuals are strictly randomly distributed in the community there is a definite and easily-calculated relationship between density and frequency.

This likewise is of little practical value for determining density from frequency values, as some earlier workers had hoped, because we now know that plants are most commonly not randomly distributed. It is important, however, in view of the biological implications of the occurrence of non-random distributions, to know the corresponding frequency for any particular density value and quadrat size in a random distribution as a basis of comparison for distributions actually found.

The nature of random distribution is not always clearly understood. In such a distribution the probability of finding an individual at a point in the area is the same for all points. FIGURE 3 shows a number of points randomly distributed within a square.

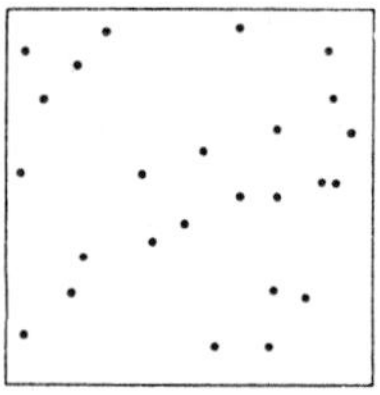

FIGURE 3. A random distribution

This figure was constructed by using two sides of the square as axes and drawing random co-ordinates from a table of random numbers. Consideration of a uniform distribution (FIGURE 2(a)) such as that of trees in an orchard, shows that the probability of finding an individual is not uniform over the area but rises at the corners of an imaginary grid. Put another way, in a random distribution the presence of one individual does not either raise or lower the probability of another occurring near by. In a uniform distribution the probability is lowered and in a clumped distribution, e.g. FIGURE 2(c), it is raised.

We may now turn to the relationship between density and frequency for a random distribution. Suppose an area A contains n individuals and is sampled by a quadrat of area a.

Let the ratio $\dfrac{A}{a} = r$

The density of individuals is $x = \dfrac{n}{A}$

The chance of finding one particular individual in a throw of the quadrat is $\dfrac{a}{A} = \dfrac{1}{r}$.

The chance of not finding that individual is therefore $1 - \dfrac{1}{r}$.

Therefore the chance of finding none of the n individuals is $\left(1 - \dfrac{1}{r}\right)^n$.

But $n = Ax = a\,\dfrac{A}{a}\,x = axr$, and $\left(1 - \dfrac{1}{r}\right)^n = \left(1 - \dfrac{1}{r}\right)^{axr}$.

Now if r is large

$$\left(1 - \frac{1}{r}\right)^r = e^{-1}\left(e = 1 + 1 + \frac{1}{2!} + \frac{1}{3!} + \cdots\right),$$

the base of natural logarithms) and the chance of the quadrat containing no individuals becomes e^{-ax}.

The chance of a quadrat containing one particular individual and no others is $\dfrac{1}{r}\left(1 - \dfrac{1}{r}\right)^{n-1}$.

Any one of the n individuals may be present in the quadrat alone so that the chance of a quadrat containing one individual is

$$n \cdot \frac{1}{r} \cdot \left(1 - \frac{1}{r}\right)^{n-1}$$

$$= n \cdot \frac{1}{r} \cdot \frac{1}{\left(1 - \dfrac{1}{r}\right)} \cdot \left(1 - \frac{1}{r}\right)^{n}$$

$$= a\,x\,r \cdot \frac{1}{r} \cdot \frac{1}{\left(1 - \dfrac{1}{r}\right)} \cdot e^{-ax}.$$

If r is large $\dfrac{1}{r}$ approaches 0 and $1 - \dfrac{1}{r}$ approaches 1 and this becomes $ax \cdot e^{-ax}$.

Similarly the chance of finding two particular individuals and no others in a quadrat throw is $\left(\dfrac{1}{r}\right)^2 \cdot \left(1 - \dfrac{1}{r}\right)^{n-2}$.

Two individuals may be selected from n individuals in $\dfrac{n(n-1)}{2!}$ ways so that the chance of obtaining two individuals in a quadrat is

$$\frac{n(n-1)}{2!} \cdot \left(\frac{1}{r}\right)^2 \cdot \left(1 - \frac{1}{r}\right)^{n-2}$$

$$= \frac{n(n-1)}{2!} \cdot \left(\frac{1}{r}\right)^2 \cdot \frac{1}{\left(1 - \dfrac{1}{r}\right)^2} \cdot \left(1 - \frac{1}{r}\right)^{n}$$

$$= \frac{axr(axr-1)}{2!} \cdot \left(\frac{1}{r}\right)^2 \cdot \frac{1}{\left(1-\frac{1}{r}\right)^2} \cdot e^{-ax}$$

$$= \frac{ax\left(ax-\frac{1}{r}\right)}{2!} \cdot \frac{1}{\left(1-\frac{1}{r}\right)^2} \cdot e^{-ax}$$

$$= \frac{(ax)^2}{2!} \cdot e^{-ax} \text{ (when } r \text{ is large)}$$

Similarly the chance of a quadrat containing three individuals can be shown to be $\frac{(ax)^3}{3!} \cdot e^{-ax}$ and in general the probabilities of a quadrat containing 0, 1, 2, 3, . . . , n, . . . individuals are given by the series

$$e^{-ax}, \; axe^{-ax}, \; \frac{(ax)^2}{2!} e^{-ax}, \; \frac{(ax)^3}{3!} e^{-ax}, \; - - - \; \frac{(ax)^n}{n!} e^{-ax}, \; - - -$$

This converging series is known as the Poisson series. The total probabilities must add up to 1 and this is easily demonstrated

$$e^{-ax} + axe^{-ax} + \frac{(ax)^2}{2!} e^{-ax} + \frac{(ax)^3}{3!} e^{-ax}, + - - -$$

$$+ \frac{(ax)^n}{n!} \cdot e^{-ax} + - - -$$

$$= e^{-ax} \left(1 + ax + \frac{(ax)^2}{2!} + \frac{(ax)^3}{3!} + - - - + \frac{(ax)^n}{n!} + - - -\right)$$

$$= e^{-ax} \cdot e^{ax}$$

$$= 1.$$

In practice the Poisson series is often more usefully expressed

$$e^{-m}, \; me^{-m}, \; \frac{m^2}{2!} e^{-m}, \; \frac{m^3}{3!} e^{-m}, \; - - - \frac{m^n}{n!} e^{-m}, \; - - -$$

where $m = ax$ is the mean number of individuals per quadrat.

Frequency is the proportion of quadrats containing at least one individual, i.e. $1 - e^{-m}$, or percentage frequency $F = 100 \, (1 - e^{-m})$. Thus density (per quadrat)

$$m = - \log_e \left(1 - \frac{F}{100}\right).$$

The relationship between density and frequency is thus logarithmic and not, as some early workers assumed, linear. This was pointed out by Svedberg (1922), Kylin (1926) and others but is still not always realized.

14

Provided that distribution of individuals is random, density may thus be obtained from frequency counts. Further, frequency for any desired quadrat size may be calculated from a determination for one size. If F_1 is the frequency at quadrat size a_1 and F_2 the frequency at quadrat size a_2,

$$F_1 = 100 \ (1 - e^{-a_1 x}) \text{ and } F_2 = 100 \ (1 - e^{-a_2 x})$$

$$\frac{F_1}{F_2} = \frac{1 - e^{-a_1 x}}{1 - e^{-a_2 x}}$$

or, better, using percentage absence

$$100 - F_1 = 100 \ e^{-a_1 x} \text{ and } 100 - F_2 = 100 \ e^{-a_2 x}$$

$$\frac{100 - F_1}{100 - F_2} = \frac{e^{-a_1 x}}{e^{-a_2 x}}$$

$$\log_e \left(\frac{100 - F_1}{100 - F_2} \right) = a_2 - a_1$$

$$\text{or } \log_{10} \left(\frac{100 - F_1}{100 - F_2} \right) = 0 \cdot 4343 \ (a_2 - a_1).$$

In view of the relative rarity of random distributions these relationships are not of great importance. It is essential to examine the type of distribution before using frequency to determine density and normally this will be more laborious than direct density determination. It might perhaps be useful to use frequency to estimate density in a few problems such as the study of minor changes in density in the same community over a period of time. In such cases even if distribution is not random it may be possible to establish an empirical relationship between density and frequency. In some cases the logarithms of percentage absence and density are proportional but density calculated from percentage absence must be multiplied by a constant to obtain the true value (Blackman, 1942). Lynch and Schumacher (1941) have shown, for seedlings in Western Pine Forest of North America, a linear relationship between the probit of frequency and logarithm of density. Goodall (1952a) points out that such relationships imply the existence of particular types of non-random distribution.

After this consideration of the nature of frequency the percentage frequency method of community description needs little further comment. We now realize that Raunkiaer's method, instead of giving a measure of the bulk of material contributed by each species, as was at first believed, is an uncertain assessment of several different characteristics, the principal value of which lies in its speed of determination. From earlier results obtained Raunkiaer

(1918, see Raunkiaer, 1934) deduced his Law of Frequencies. If the total number of species in a community is divided into the following five classes

A   0–20 per cent frequency
B   21–40   „   „
C   41–60   „   „
D   61–80   „   „
E   81–100 „   „

the law of distribution of frequencies states that A> B> C$\gtreqless$D <E. This is shown graphically in FIGURE 4.

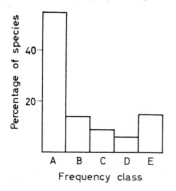

FIGURE 4. 'Law of Frequencies.' Total data for a number of Scandinavian plant communities. Frequency classes: *A*, 1–20 per cent; *B*, 21–40 per cent; *C*, 41–60 per cent; *D*, 61–80 per cent; *E*, 71–100 per cent (data from Raunkiaer, 1934)

The general fall in the first three or four frequency classes accords with the experience of field botanists that there are more rare species than common ones, whatever the area under consideration. The rise in the fifth class was unexpected and various explanations of it have been advanced. It has played a notorious part in some attempts at definition of plant associations. The explanations advanced are largely invalidated, as Kylin (1926) pointed out, by the fact that the shape of the resultant curve is largely determined by the method of sampling (see also Ashby, 1935). These will not be discussed here. TABLE 2 shows the densities corresponding to the boundaries of Raunkiaer's frequency classes. It will be seen that the third class covers a range of density of 0·405 and the fourth class of 0·693 but the fifth class includes all densities from 1·609 to the maximum found. This maximum value is limited theoretically by the size of an individual, which is small compared with the size of quadrats commonly used. (Many workers, following Raunkiaer, have used 1/10 square metre quadrats, i.e. a square of side 31·6 cm.) In practice densities many times 1·609 are commonly found. The fifth class thus covers a range of density exceeding that of the

remaining classes put together and a correspondingly large number
of species fall into it, the increase in number due to greater range
more than counterbalancing the fall off in number of species with

TABLE 2

DENSITY CORRESPONDING TO CERTAIN FREQUENCIES IN RANDOM
DISTRIBUTIONS (see also Appendix B, Table 6)

Frequency (%)	Density (number per quadrat)	Range of density included in Raunkiaer's frequency classes
0	0	0·223
20	0·223	0·288
40	0·511	0·405
60	0·916	0·693
80	1·609	(1·609 to maximum
100	∞	present)

increasing density. If distribution is more uniform than random
expectation, the density limits of the frequency classes will be
lowered and the rise in the fifth class will be still more pronounced.
If species are aggregated, the limits will be increased but the effect
will still be found unless the aggregation is very pronounced.
Preston (1948), discussing the relative proportion of rare and
common species, shows that the results expressed as Raunkiaer's
Law of Frequency follow from an assumption that the numbers of
species of different degrees of rarity and commonness fall on a
lognormal curve. (Here 'rare' and 'common' refer to number of
individuals per association or other statistical universe.) The
results may also follow from other assumptions of the relation
between numbers of species of differing degrees of rarity (cf.
Williams, 1950). It is clear that the exact form of the curve of
frequencies depends on a complex of factors, including the make-up
of the population in terms of relative numbers of individuals of
different species, pattern assumed by individuals of different species,
and size, and, possibly (Williams, 1950), number of quadrats used.

The relationship between shoot frequency and other measures of
vegetation has been little considered. We might expect that there
would be a similar relation between shoot frequency and cover to
that between rooted frequency and density. Blackman (1935) has
shown a close correlation between cover and percentage absence for
Trifolium repens in grassland communities. It is evident that cover

17

cannot be randomly distributed but always must show aggregation with areas of high cover at least the size of individual plants. A theoretical relationship between cover and shoot frequency for random distribution of cover would therefore be of no practical value. Blackman's observations show that an empirical relationship may be obtained in at least some cases, which is of similar utility in intensive studies to the empirical relationship obtainable between density and rooted frequency. Aberdeen (1954, 1958), following a suggestion of Archibald (1952), has drawn together the two concepts of rooted and of shoot frequency by considering frequency of plant units of radius r and density d in a circular quadrat of radius R. Assuming random distribution, shoot frequency $F = 100$ $(1 - e^{-\pi(R+r)^2 d})$ and estimates of both density and plant size may be obtained from frequency data for two quadrat sizes. He has applied a similar argument to volume samples for estimations of soil fungi, when $F = 100$ $(1 - e^{-\frac{4}{3}\pi(R+r)^3 d})$ (Aberdeen 1955). As in other frequency approaches, the utility of the relationship is limited by the general departure from randomness found in the field.

In intensive studies of particular species it is often necessary to obtain some measure of the performance of the species under different conditions. The possible measures are numerous and the most suitable one to use must depend on the growth form of the species and the particular aspect of its behaviour under investigation. Direct measures of size, or number of parts, or estimates of yield, may all be appropriate under different conditions. Cover repetition, the mean number of hits by pins, at points where at least one hit is made, is basically a measure of performance, as it reflects the number of layers of foliage produced by an individual. It is not necessary to enlarge on the possible measures here, but it is perhaps worth emphasizing that less obvious measures may be valuable indicators of performance. Thus Phillips (1954b) found that the ratio of solid tip to grooved blade in the leaves of *Eriophorum angustifolium* reflects the rates of production of leaf primordia relative to rate of leaf elongation, and used the ratio to correlate these growth rates with environmental conditions. A possible future development is the use of multivariate analysis to extract from a variety of measures of performance the underlying *factors**. This approach has apparently been used only in agricultural and horticultural contexts. Thus Pearce (1958) used four measures of size in

---

* *Factor* is here used in a sense, well established in the context of multivariate analysis, different from the usual ecological usage. It refers essentially to a set of correlations between different variates. Such a factor may be judged to be more or less controlled by one or more ecological factors but is not necessarily so controlled.

apple trees, circumference at four years and at grubbing, tree weight at grubbing and extension growth in the first four years. Data were available for a number of trees and he was able to show that a large part of the variability in these measures could be accounted for in terms of two factors. One was related to vigour, the tendency to make growth in all its forms, the other to establishment, the tendency for some trees to do better and others to do less well than their early promise. Establishment was in turn related to soil conditions but vigour was largely hereditarily determined. Similar approaches could profitably be used in autecological studies to give a more precise meaning to the concept of performance.

CHAPTER 2

SAMPLING AND COMPARISON

THE value of quantitative data on the composition of vegetation depends on the sampling procedure used to obtain it. Since collection of quantitative data in the field is at best a time-consuming task, it is imperative that the samples taken should be such as give the maximum amount of information in return for the effort and time involved. With few exceptions the object in making quantitative estimates of vegetation will fall into one or other of three categories: (a) an estimate of the overall composition of the vegetation within certain boundaries, with a view to comparison with other areas or with the same area at another time, (b) the investigation of variation within the area, or (c) correlation of vegetation differences with differences in one or more habitat factors. This division is not absolute but it is worth making, because, for instance, the most useful procedure for sampling for overall composition may not be the most satisfactory for examination of variation within the area. Sometimes a compromise is possible whereby more than one objective may be attained, although perhaps at a lower level of precision, by suitable sampling procedures. This chapter is mainly concerned with the first category of sampling overall composition, though some of the general principles to be mentioned are applicable to all sampling. For a more comprehensive account of the theory of sampling, reference may be made to accounts such as those of Sampford (1962) and Cochran (1963).

There must inevitably be an element of subjectivity in sampling procedure because the boundaries within which a set of samples is taken are fixed by the ecologist on the basis of his judgement of what can suitably be described as one unit for the purpose in hand. This handicap is less serious in agricultural ecology, where man-made influences play a large part in determining vegetation and have commonly been applied uniformly over defined areas. But even when the grassland worker uses fields as his units he is depending on his own judgement that the influence of identical management over the field overrides any diversifying effect of soil or microtopography. This subjective element cannot be eliminated but steps can be taken to minimize its effects. Firstly, parts of an area may be sampled separately if there is doubt as to its homogeneity. The data can be lumped to give a single value if comparison of the results for the different parts shows them to be sufficiently alike by whatever standards seem appropriate to the particular problem, and here again subjective judgement is involved. Secondly, the sampling may be carried out in such a way that information may be obtained on homogeneity and that some division of the data into separate portions corresponding to parts of the area may be possible after sampling has been completed. Common sense must be used in applying these precautions. Clearly any considerable fragmentation of an area for descriptive purposes cannot be carried out without so reducing the information on each part that it has little value by itself, or else so increasing the number of samples that the labour involved is out of proportion to the information to be obtained. Sampling procedure must always be related to the importance of the information sought to the problem under investigation, and to the degree of precision necessary.

Within an area selected and defined on this basis samples might be placed in three ways: by selecting sites considered typical of the area as a whole, by placing samples randomly or by placing them systematically in some regular pattern, or some combination of these methods. Selection of typical sites for samples is clearly inappropriate to a quantitative approach, as their choice is dependent on the observer's preconceived ideas of the character of the vegetation, and data from such samples cannot be considered an unbiased estimate of the vegetation of the area. The alternatives of random or systematic sampling merit rather more consideration. If samples are taken at random there is available not only an estimate of the mean value of the density, or whatever measure may be used, but also an estimate of the precision of this mean. This, normally expressed as the standard error of the mean, allows a statement that, within any desired probability, the true value lies within a certain range. Thus, suppose the mean density from sixty quadrat samples is 11·5 individuals per unit area with a standard error of 1·5. For 59 degrees of freedom the 1 per cent level of t is (from tables) approximately 2·7, so that there is a 1 per cent probability of obtaining a deviation of $2·7 \times 1·5 = 4·05$, or greater, from the true value by chance, or 99 per cent probability that the true value lies within the range $11·5 \pm 4·05$. Such an estimate of the precision of the mean is desirable even if the data for one area are considered alone. If we wish to compare the densities in two different areas it becomes essential; only if the accuracy of the two estimates is known

21

is it possible to assess the likelihood of getting by chance such divergent means as those observed in two different sets of samples from one population. Suppose a density of 10 individuals per unit area is observed in one community and 19 in another. An observer ignorant of the implications of sampling might say that since one value is nearly twice the other there must be a real difference in density. To do so is quite unjustified. Without knowledge of the precision of the two figures no conclusion can be drawn as to whether there is any real difference in density between the two communities. If, however, we know the standard errors of the two means, then a statement of the probability of the difference arising by chance is possible. Suppose the standard error of the first mean is 3 and that of the second is 4, then the standard error of their difference is $\sqrt{(3^2 + 4^2)} = 5$. The observed difference is 9 and $t = \dfrac{9}{5} = 1 \cdot 8$. Since the 5 per cent point for t based on a reasonably large number of observations is about 2, the probability of getting such a large difference by chance, even if the two populations were identical, would be more than 5 per cent, and we can scarcely consider this adequate evidence of real difference. It must be emphasized that the final judgement is subjective. A statistical test can indicate only the probability, on a given hypothesis, of a particular kind of result arising by chance.

If sampling is systematic an estimate of the mean is available which may, in some circumstances at least, deviate less from the true value than that given by random samples, but there is no indication of its precision and no possibility of assessing the significance of its difference from the mean in another area. In spite of the latter disadvantage systematic sampling has been preferred by many workers on the grounds that it is more representative of variations over the area and hence likely to give a better estimate than random samples, and that it is easier to carry out in the field. Several comparisons of random and systematic sampling have been made, all apparently on forest trees. Hasel (1938) and Finney (1948) for timber volume, and Bourdeau (1953) for density and basal area, found that any gain in accuracy from systematic sampling was slight. Finney (1950) found that, if there was a strongly marked pattern of variation over the area, there was little advantage in systematic sampling, and, if the pattern was periodic, the values obtained by systematic sampling were likely to be less accurate than those from random. In these investigations the variance of the observed systematic samples was used as a measure of their accuracy, but it must be emphasized that the variance is here being used as a measure of conformity between successive samples over the same area and cannot be justified as a basis of tests of significance between samples taken over different areas. It is evident then that, apart from the desirability of being able to make tests of significance, whether systematic sampling gives a more accurate value than random depends on the pattern of variability within the area.

If systematic sampling is desired for a particular investigation, it is important that the pattern of sampling adopted should be such as to give uniform representation over the area. Otherwise the advantage that may result from its use will be lost. This point is not always appreciated. For example, Brown (1954) quotes an investigation in which plots, presumably rectangular, were examined by points placed at equal distances along the two diagonals of the plot and along lines joining the midpoints of the opposite sides. This results in a much greater intensity of sampling towards the centre of the plot; half the samples are in fact taken from a quarter of the area.

We have so far considered only entirely systematic or entirely random sampling. Of the two it is clear that, in most circumstances, random sampling is the better if overall information on the composition is desired, as in many types of applied work and in the classification of vegetation. Systematic sampling, as will become clear in Chapter 3, is, on the other hand, more advantageous if interest centres on variability within the area. The choice is, however, not necessarily between entirely systematic and entirely random samples. The essential feature of random sampling, which allows the observed variance of the data to be used as the basis of tests of significance, is that any point within the area has an equal chance of being represented in the samples. Randomization may be restricted in any way such that this is still true. Thus, if the area is divided up into a number of blocks of the same size, even if they are not of the same shape, and the same number of samples taken at random within each block the condition will be satisfied. Provided the blocks are not comparable in size to the scale of variability within the area, this will result in no loss of precision compared with random sampling, even if the variability present is periodic, and will give an increase in precision comparable to that from entirely systematic sampling under conditions where systematic sampling would do so. Theoretically the greatest advantage will be obtained by subdivision as far as possible, so that only one sample is taken from each block. In practice a balance must be struck between this and the increasing labour of laying down large numbers of boundary lines. This increased precision of restricted as against unrestricted random sampling has been demonstrated by a number of workers, e.g. Pechanec and Stewart (1940) for yield, Bourdeau (1953) for density and basal area, and Goodall (1952b) for cover. Division into blocks has a further advantage. After the samples have been taken it may become apparent that the area should have been regarded as two or more distinct units. If so, it is possible to assign observations to one or other of the units and obtain a mean value for each of the latter. Indeed, especially if the blocks are all of the same shape, the arrangement facilitates judgement on this by allowing one portion of the area to be tested against another.

Random placing of a quadrat or other sample is not as straightforward as is sometimes thought. It is not uncommonly believed to be sufficient to walk over an area, throwing the quadrat over one shoulder or throwing with the eyes shut or in some other way that apparently avoids any deliberate choice of the exact position of the quadrat. It is an instructive exercise to do this, at the same time plotting the position of each quadrat. If the positions are tested for randomness (e.g. by the method of grid analysis described in Chapter 3), it will almost always be found that the samples are not randomly distributed over the area. Commonly some portions of the area will have been omitted and those that have been sampled will show too regular a pattern of samples. Even if small plots a few feet square only are being sampled and care is taken to throw the quadrats over a larger area, ignoring those falling outside the boundaries, it is difficult to avoid a consistent over-representation of the centre of the plots (cf. Greig-Smith, 1952a). The extra trouble of some more objective method of randomizing is, therefore, generally well worth while. This is most readily done by laying down two lines at right angles as axes (which for rectangular plots are most conveniently two adjacent sides of the plot) and using a pair of random numbers as co-ordinates to position each sample. Measurement of the distances from the axes to sample positions need not be exact—pacing is quite sufficient in large plots. Selections of random numbers are readily available (e.g. Fisher and Yates, 1943; Snedecor, 1946).

Detailed examination of an area may be intended as a basis for study of changes in its vegetation over a period rather than for a comparison with the vegetation of other areas. The usual practice is to take a suitable series of samples at the beginning of the period, preferably by some scheme of restricted randomization, and another series at the end of the period, using the same sampling procedure. Goodall (1952b) has pointed out that if the same samples (point quadrats in the case he is considering) are examined at the beginning and end of the period, a large proportion of the sampling error of individual points is removed from the estimate of change. He quotes data for percentage cover changes over a year obtained from permanent point quadrats and from point quadrats located afresh at the second examination. Out of data for eight species and percentage bare ground, in all cases except one the variance of the difference is less for permanent points (in most cases markedly so); for one species it is less than one-fifth the corresponding value for independent points. It is important that the marking of points is done in such a way that it does not interfere with the vegetation; a peg at the point of sampling is undesirable—fixing by measurement from two pegs a little distance away is better. Goodall notes a difficulty that may arise with fixed samples if several successive records are to be taken. This is the possibility that successive changes at one point may be correlated. It is not certain how likely this is in natural vegetation, but the possibility clearly has to be taken into account. The difficulty may be overcome by recording at the end of the first period a second set of samples which are used to estimate changes in the second period, and so on, i.e. samples A are enumerated on the first occasion, samples A and B on the second, samples B and C on the third, etc.

One further general consideration on sampling procedure remains before turning to the different measures of vegetation. In large-scale projects a number of observers may be involved. The differing bias of different observers estimating vegetation subjectively, referred to in Chapter 1, is well known. That observers enumerating objective measures of vegetation may also exhibit personal bias is more commonly overlooked. With most measures and types of sampling it is likely to be small and unlikely to be significant unless small differences are important in the investigation. Ellison (1942) found considerable differences between estimates of cover (from 17 to 22 per cent, using 2,400 points) of the grass *Buchloë dactyloides* by different observers using a frame of pins in the same position, but each dropping the pins afresh. Goodall (1952b) concluded from similar experiments, but with the pins remaining in position for all observers, that part of the error was due to movement of the pins, and that observers experienced in the method could obtain considerably better agreement. However, even under these conditions, although observers A and B agreed in their estimates for all seven species examined, observer C significantly exceeded the estimates of A and B for two species and fell below them for a third. In cover estimates the personal factor derives from judgement whether a pin is or is not touched. In other measures similar effects can be expected, e.g. in deciding whether an individual on the boundary of a quadrat does or does not lie within it. There is likewise scope for personal judgement in deciding whether an axillary bud starting growth has developed far enough to be considered a new tiller. In a large-scale investigation employing a number of observers, particularly if it is prolonged, it is worth while for the observers to check their assessments against one another on the same samples from time to time. If persistent differences of the same sign appear, it may be feasible to calculate correcting factors to make their estimates strictly comparable. If several observers are working together on the same areas, it is certainly worth while to divide the sampling among them in such a way that personal error is included in the sampling error and not in between-plot or between-area differences, e.g. if two observers are recording a number of plots, they should each record half the samples on each plot rather than each recording all the samples on half the plots. It is scarcely necessary to emphasize the importance of establishing beforehand the exact details of sampling procedure to be adopted in any investigation involving several observers, as even slight deviations in method between observers may invalidate comparisons of their data.

Turning to considerations peculiar to different measures we may consider first density. Three decisions have to be made before sampling is begun. These concern the size and shape of quadrat and the number of samples to be used, considerations which are not necessarily independent of one another. If the individuals counted are randomly distributed over the area, i.e. in this context, if the numbers per quadrat of the size and shape used fall on a Poisson series, the accuracy of the estimate obtained can readily be shown to depend only on the number of individuals counted. Suppose x individuals are enumerated in n quadrats. Then the mean number per quadrat is $\dfrac{x}{n}$ and, since the variance of a Poisson series is equal to its mean, the variance of a single quadrat is also $\dfrac{x}{n}$.

Variance of the mean of n quadrats then $= \dfrac{1}{n}\left(\dfrac{x}{n}\right)$.

Standard error of the mean $= \dfrac{\sqrt{x}}{n}$.

Ratio of standard error of mean to mean $= \dfrac{\sqrt{x}}{n} \cdot \dfrac{n}{x} = \dfrac{1}{\sqrt{x}}$.

Thus the standard error of the estimate obtained will be the same for the same number of individuals counted, whether many small quadrats or few large ones are used, or even a single sample only counted. That this relationship applies even to a single quadrat is not perhaps immediately obvious, but it follows from the fact that one single count of x random individuals in a quadrat can be regarded as one sample from a whole series of such samples. It has therefore a variance equal to its mean (x) and a standard error $\sqrt{x}$. With a single quadrat the estimate of accuracy is in fact approximate only as the observed value x may deviate considerably from the mean of the series from which it is drawn. Indeed the ratio $\dfrac{1}{\sqrt{x}}$ can only be used for reasonably large numbers of quadrats. If the individuals of the species under consideration were randomly distributed, any convenient size of quadrat could be used and counts repeated with it until sufficient individuals had been enumerated to give the required accuracy.

The theoretical relationship for density data between the mean and its standard error is in practice of little assistance in deciding the number of quadrats necessary. In the field individuals are almost always found not to be randomly distributed but to show contagious distribution. In the present connection the important feature of this is that the variance is greater than the mean. Very rarely individuals may be regularly distributed, with variance less than the mean. Thus the calculated value of $\dfrac{1}{\sqrt{x}}$ for the ratio of standard error of mean to mean for x individuals counted is of very limited use as a guide to the number of samples that should be used. At the most it can be predicted that the accuracy attained is most unlikely to be as great as that indicated by the theoretical value and may be very much less.

When individuals are not randomly distributed, not only is the variance not equal to the mean but, not surprisingly, it is found not to be proportional to it either. Under these conditions the size of quadrat used and possibly its shape also will affect the accuracy of the estimate of density obtained. TABLE 3 shows data for mean and variance of density obtained by two different sizes of quadrat on the same area, and illustrates the kind of difference commonly found in the field. Here a reduced number of a larger size of quadrat gave an increase in the standard error expressed as a proportion of the mean. This is the commonest result to find, at least within the range of quadrat sizes that are practicable in the field, but it is

27

possible to obtain the reverse effect. Fuller examination of these alternatives must be postponed to the discussion of pattern within communities in Chapter 3. We may anticipate briefly this discussion by noting that if a non-random population is sampled by quadrats of a size very much smaller than the average size of patches of

DENSITY ESTIMATES OF SHOOTS OF *Mercurialis perennis* MADE IN THE SAME AREA WITH TWO DIFFERENT SIZES OF QUADRAT

	Quadrat size	
	625 sq. cm	2,500 sq. cm
Mean number per quadrat	0·13	0·35
Variance	0·2343	1·1432
Variance of mean for same total area sampled (200 quadrats of 625 sq. cm / 50 quadrats of 2,500 sq. cm)	0·00117	0·02286
Standard error of mean	0·0342	0·1512
Standard error of mean/mean	0·263	0·432
Estimate of density per sq. m	2·08	1·40
Standard error of estimate	0·547	0·605

individuals, then the variance of an observation will not be much, if any, greater than the mean. As quadrat size increases and approaches the size of the patches variance relative to the mean will rise sharply. If the patches are regular, it will then fall off again, ultimately reaching, or even falling below, the mean. If, however, the patches are themselves randomly or contagiously distributed the high variance will be maintained. Unfortunately it is rarely possible to determine by inspection whether the patches are regularly arranged, especially as regular arrangement most usually occurs as a mosaic of areas of slightly different density, which, being contiguous, must be regular in the sense in which the term is used here. The position may be and often is complicated further by the occurrence of heterogeneity on several different scales in the area. Any exact investigation of pattern is far more time-consuming than direct determination of density, so that there can be no question of investigating pattern before making a density determination. The safest procedure thus appears to be to use for density determinations the smallest quadrat that is practical or desirable on other grounds.

It has long been the custom to use a square sampling area and many ecologists have probably never given thought to its efficiency compared with other possible shapes. Clapham (1932) showed for one particular case that the variance between rectangular strips was markedly less than between squares, and that the variance was least, i.e. the efficiency was greatest, if the strips were orientated at right angles to the boundaries between any obviously different parts of the area sampled. This conclusion has been confirmed in various communities (see, for example, Bormann, 1953). This could not apply to perfectly random distributions where the dependence of standard error only on number counted would hold, and its general occurrence is in itself evidence of the widespread non-randomness of vegetation. It is easy to see why it should be so in non-random populations; a more elongated unit is more likely to include portions of more than one of the density phases which make up the population. Clapham also pointed out the greater ease in the field of working with strips than with squares, in avoiding trampling on part of the sample area while examining another part, and in dividing it up for counting. Too great an elongation of the strip, however, carries with it disadvantages of increased edge-effect similar to those found with very small square quadrats, and the exact shape must be determined with the growth-form of the species being counted in mind.*

The minimizing of variance is one consideration determining the size of quadrat to be used, but there are others also. The first is a practical one; the smaller the quadrat the greater the length of quadrat boundary per unit area and consequently the greater the chance of significant edge-effects due to the observer consistently including individuals that ought to be excluded or vice versa. For this reason alone, particularly if individuals of the species under consideration have a large area or ill-defined boundary at ground level, it is advisable that the quadrat should not be too small.

The second consideration is less obvious; like the last it applies equally to random and non-random populations. Consider first a random population in which the occurrence of individuals in a quadrat follows a Poisson distribution. FIGURES 5–8 show the frequency distribution of samples with 0, 1, 2, etc., individuals per quadrat for different means. It will be seen that for low values of the mean the distribution curve is highly asymmetric. Now the usual statistical procedures for the comparison of means are based on the assumption that the samples being compared are drawn from populations showing normal distribution and in which the variances are independent of the means. Poisson distributions

---

* Myers and Chapman (1953) claimed to have found that rectangles showed *greater* variance than squares of the same area in *Leptospermum* scrub, but their analysis of variance in fact demonstrates the greater variance of squares.

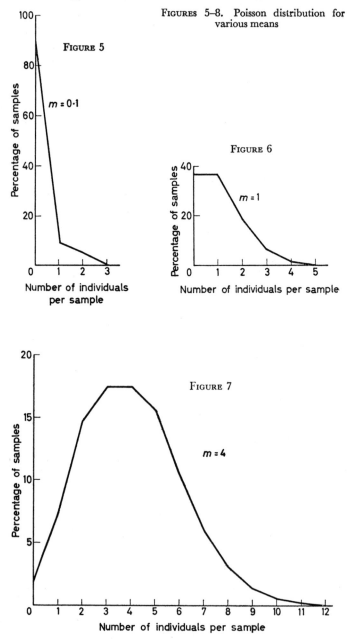

FIGURES 5–8. Poisson distribution for various means

FIGURE 5

$m = 0.1$

Percentage of samples

Number of individuals per sample

FIGURE 6

$m = 1$

Percentage of samples

Number of individuals per sample

FIGURE 7

$m = 4$

Percentage of samples

Number of individuals per sample

satisfy neither of these conditions. As the mean increases, the form of the distribution curve approaches that of the normal (cf. the Poisson distribution for a mean of 9 with the normal distribution of mean and variance both 9 in FIGURE 8) but the condition of independence of variance is still not satisfied. These difficulties can be overcome, provided the mean is sufficiently high (appreciably greater than 1), by making a suitable transformation which should always be done before comparisons between means of random populations are tested for significance. For a Poisson distribution this takes the form of substituting for each quadrat reading its

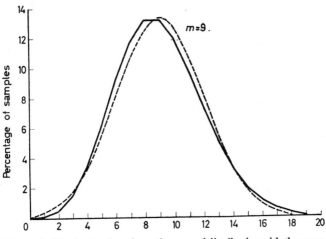

FIGURE 8. The broken line shows the normal distribution with the same mean and variance

square root. If the mean number is less than 10 this tends to overcorrect and a more satisfactory transformation is $\sqrt{}$(observed number + 0·5); this should be used if any of the values being compared are less than 10.

If the distribution of individuals is contagious the asymmetry of the distribution curve for low means will be still more pronounced, but the dependence of the variance on the mean is likely to be less pronounced, on account of the varying factors affecting pattern in different areas. If means are made reasonably large by a suitable size of quadrat, the degree of asymmetry will usually be insufficient seriously to affect tests of significance. In deciding quadrat size, therefore, reasonable symmetry of the distribution curve should be aimed at. In practice this may be interpreted as selecting a size which will not give more blank quadrats than quadrats with one individual. Before making comparisons the data should be scrutinized to see if any sets are very asymmetric or if the variances of the different sets tend to be proportional to the means. If so the square root transformation should be used.

The factors affecting the accuracy of the mean in determinations of density in the field are thus varied and complex and the theoretical relationship for random populations between number of individuals counted and accuracy is of little practical value. It remains broadly true, however, that the larger is the number counted the greater is the accuracy of the resulting mean. It is possible to calculate the variance from time to time as sampling proceeds but this course scarcely commends itself and is unlikely to be adopted unless a certain minimum level of accuracy is essential to the investigation. A rough indication may be obtained while the sampling is in progress whether further samples will greatly increase the accuracy of the mean by calculating successive means. The mean of, say, the first five, ten, fifteen, twenty, etc., observations is calculated and plotted against the number of observations. The mean will at first oscillate violently but gradually the oscillations will become less as the sample size increases. The graph can be judged in a subjective way only but does give the negative indication that if large oscillations are still occurring the sampling should be continued. FIGURE 9 shows an example of such a graph together with the percentage standard error of the mean. The theoretical value of the standard error, assuming individuals to be randomly distributed, is also shown and emphasizes the lower degree of accuracy usually found in field data. The low standard error found for the first ten observations results from their chancing to be unduly uniform. The mean of the complete set of 100 observations has a standard error of 8·14 per cent, compared with a calculated value of 5·31 per cent.

Another useful approximate check in the field is derived from the relationship between range and standard error. If successive sets of samples are taken from a normal population, the mean range of the sets is a definite multiple of the standard error of the population, e.g. samples of five observations have, on the average, a range 2·326 times the standard error. Thus the percentage standard error of the mean of any total number of samples, or alternatively, the number of samples necessary to give any desired degree of accuracy, may be estimated from the mean range. Appendix B, TABLE 2 shows the value of the constant for different sizes of sample. In the data shown in FIGURE 9 the ranges of the successive groups of five observations were 5, 6, 2, 5, 11, 5, 13, 4, 4, 5, 7, 12, 8, 6, 6, 3, 7, 11,

6, 5. The mean value is 6·55 and the estimate of the standard error is $\dfrac{6·55}{2·326} = 2·82$. From this the standard error of the mean of the 100 observations is estimated as 0·282, or 7·93 per cent, compared with the observed value of 8·14 per cent.

When concurrent determinations of density for more than one species are desired, the optimum size of quadrat and the number of quadrats necessary are often not the same for different species. A compromise must then be made in deciding the size and possibly

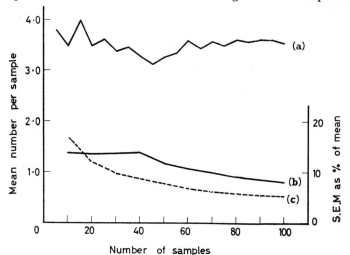

FIGURE 9. Number of individuals of *Endymion nonscriptus* in 100 random quadrats (10 cm square). (a) Mean of first 5, 10, 15, . . . 100 samples. (b) Observed and (c) calculated (for random distribution) standard error of mean, expressed as percentage of mean, for the first 10, 20, 30, . . . 100 samples

even the shape of quadrat to be used. There is no reason, however, why the number of quadrats used for different species should be the same. There will clearly be a saving of labour if counting of the more abundant species is stopped after the required number of quadrats and only the rarer species recorded in the later samples.

Sampling yield in its various forms is more straightforward than sampling density, as the data obtained will usually show an approximately normal distribution with variance independent of the mean and will vary continuously (in the mathematical sense, that a reading can take any one of a large number of values. Beware, however, of very small ranges with a low degree of accuracy in weighing or measuring, e.g. if yields range from 0 to 3 grammes and it is possible only to weigh to within 0·5 gramme, only seven values are possible). Since distribution of data is normal, no prediction can be made of the number of samples necessary for a given degree of accuracy in the mean, unless there is information on the variance found for the same species in other closely similar communities. Otherwise the number to be used must be decided from the data as sampling proceeds. It is to be expected that the smaller the sampling unit the lower will be the variance of the mean for comparable areas sampled. This indicates the use of a small quadrat, but, as with density, edge-effects become more prominent as the quadrat is reduced in size. A balance must thus be struck between the lesser edge-effect (and the convenience of sampling) of larger quadrats and the greater efficiency per unit area of small quadrats. No guide but experience and trial of different sizes can be offered towards effecting this compromise. Small sample size may give pronouncedly skewed distribution curves with some yield data but this may often be corrected by logarithmic transformation (replacement of a reading x by $\log x$) as Blackman (1935) suggests.

Comparison of mean density or yield in two areas may be made by the t test, or joint comparison of several areas by the usual analysis of variance procedure, after transformation of the data, if this is necessary. It is important to realize the exact nature of the null hypothesis when these procedures are used. They are designed to test whether two or more samples can be regarded as *drawn from the same normal population*. This implies not only that they have the same mean but also the same variance. In most experimental work in botany the conditions of experiments are such that the variance remains the same for different treatments, and there is a tendency to regard the t test as testing difference of mean only. In the measurement of vegetation there is no reason to suppose that the variance will be the same in different areas, and this complicates the interpretation of the t test as usually applied, as a large value of t may result either from a difference of means or a difference of variance or both.

Full discussion of the testing of the difference of means (which is generally the main objective) when variance is heterogeneous is beyond the scope of this book but some account is given below of approximate treatments sufficient for most ecological work. An introduction to the subject, with references to more detailed accounts, is given by Snedecor (1946, § 4.6). In making a t test, if there is no evidence of heterogeneity of variance, a pooled estimate of the variance of an observation is obtained from the two sets of data. From this an estimate of the variance of a mean is obtained by dividing by the number of samples contributing to the mean.

The variance of the difference of two means is the sum of the variances of the two means. Thus, for two sets of data

$$x'_1, x'_2, x'_3, x'_4, - - - x'_{n'}, \text{ and } x_1'', x_2'', x_3'', x_4'', - - - x''_{n''}.$$

The pooled estimate of variance is

$$V = \frac{S (x' - \bar{x}')^2 + S (x'' - \bar{x}'')^2}{n' + n'' - 2}.$$

Variance of difference of means $\dfrac{V}{n'} + \dfrac{V}{n''} = V \left(\dfrac{1}{n'} + \dfrac{1}{n''} \right).$

Standard error of difference of means $\sqrt{\left\{ V \left(\dfrac{1}{n'} + \dfrac{1}{n''} \right) \right\}}$ or, if

$n' = n''$, standard error of difference is

$$\sqrt{\left\{ \frac{2}{n} \left[\frac{S (x' - \bar{x}')^2 + S (x'' - \bar{x}'')^2}{2n - 2} \right] \right\}}$$

$$= \sqrt{\left[\frac{1}{n} \cdot \frac{S (x' - \bar{x}')^2 + S (x'' - \bar{x}'')^2}{n - 1} \right]}.$$

Note that if the number of samples in the sets is the same, the same value is obtained by calculating the variance of the two sets separately and adding the resulting variances of the two means, thus:

Variance of first mean $\dfrac{1}{n} \cdot \dfrac{S (x' - \bar{x}')^2}{n - 1}.$

Variance of second mean $\dfrac{1}{n} \cdot \dfrac{S (x'' - \bar{x}'')^2}{n - 1}.$

Variance of difference of means $\dfrac{1}{n} \cdot \dfrac{S (x' - \bar{x}')^2 + S (x'' - \bar{x}'')^2}{n - 1}.$

The ratio of the observed difference of means to its standard error is referred to the table of t with $(n' + n'' - 2)$ degrees of freedom.

If the variances of the sets are markedly different, and it is desired to test the significance of difference of means, this procedure is modified (Snedecor, *loc. cit.*).

If the number of samples is the same in the two sets, t is referred to the table with $n - 1$ instead of $2n - 2$ degrees of freedom. The probability obtained is accurate to two places of decimals.

If the number of samples is different in the two sets, an approximate method due to Cochran and Cox (1944) may be used. The variances of the two means are calculated separately and the variance of the difference obtained by addition. To obtain the value of t corresponding to any probability level the values of t at that probability for the numbers of degrees of freedom in the two sets are

35

combined in a weighted mean, the respective weightings being the variances of the corresponding means. For example, suppose one mean has a variance of 0·14 and is derived from 10 samples and another has a variance of 0·35 and is derived from 25 samples. The 5 per cent value of t for 9 degrees of freedom is 2·262 and for 24 degrees of freedom is 2·064. The 5 per cent value of t for the comparison of the two means is then

$$\frac{0·14 \times 2·262 + 0·35 \times 2·064}{0·14 + 0·35} = 2·121.$$

In practice, if both means are based on a large number of degrees of freedom, as is often the case at least for density and yield data, it will rarely be necessary to calculate the weighted mean of t as the two values will be so close that the observed t will be likely to be either above or below both values.

The way in which difference in variance may mislead in comparison of means, if the usual t test based on a pooled estimate of variance is used, may be illustrated by a hypothetical example.

In community A 100 samples showed the following data for density

$$Sx = 960, \ n = 100, \ \bar{x} = 9·6, \ Sx^2 = 9,711, \ S(x - \bar{x})^2$$

$$= 9,711 - \frac{960^2}{100} = 495,$$

$$V_x = \frac{495}{99} = 5, \ V_{\bar{x}} = \frac{5}{100} = 0·05$$

In community B 50 samples showed the following data

$$Sx = 570, \ n = 50, \ \bar{x} = 11·4, \ Sx^2 = 8,948, \ S(x - \bar{x})^2$$

$$= 8,948 - \frac{570^2}{50} = 2,450$$

$$V_x = \frac{2,450}{49} = 50, \ V_{\bar{x}} = \frac{50}{50} = 1.00$$

The pooled estimate of variance for the two sets is

$$\frac{495 + 2,450}{148} = 19·899$$

Variance of difference of means $19·899 \left(\dfrac{1}{100} + \dfrac{1}{50}\right) = 0·5970$

Standard error of difference of means $\sqrt{0·5970} = 0·773$

$t = \dfrac{11·4 - 9·6}{0·773} = 2·33$ with 148 degrees of freedom and probability less than 5 per cent.

On the basis of this test a real difference between the two means might be suspected. Examination of the two means and their standard errors, however, indicates that both might be estimates of a true value of, say, 9·7, and the absence of any real difference is confirmed by the modified test

Variance of mean for A 0·05

Variance of mean for B 1·00

Variance of differences of means 0·05 + 1·00 = 1·05

Standard error of differences of means $\sqrt{1\cdot05} = 1\cdot025$

$$t = \frac{1\cdot8}{1\cdot025} = 1\cdot76$$

t for 49 degrees of freedom at 5 per cent level of probability is about 2·01 and for 99 degrees of freedom about 1·99. There is thus no need to calculate the weighted value as the observed t is below both values and there is no indication of difference of means. The significant value of t by the normal test thus derives from a difference in variance and not one in mean.

Difference between the variances of density in different areas is unlikely to be of direct interest, beyond its effect on testing differences of means. It may sometimes be important where measures of individuals are used as indicators of performance, and the testing of significance of difference of variance is therefore worth mentioning. If two estimates only are involved, the ratio of the larger to the smaller may be referred to tables of variance ratio (F), entered with the larger number of degrees of freedom at n_1 of the table. Since the larger variance is always made the numerator of the ratio the tabular probabilities must be doubled. If several variances are to be compared, Bartlett's test is available (Bartlett, 1937). Reference may be made to Snedecor (1946, § 10.13) for an account of this test.

In sampling frequency the problems are rather different from those met with in density and yield estimates. A figure for frequency expresses the proportion of samples in which a species occurs. Hence successive sets of samples from the same population will be binomially distributed. This applies whatever the pattern of individuals within the community sampled, provided that the samples are completely random. For such a set of random samples the variance of an observed number of occupied quadrats is estimated directly from the appropriate binomial series, and is npq for a binomial series $(p + q)^n$ where n is the number of quadrats in the set, p the chance of the species being present in any one quadrat and $q = 1 - p$. Thus if 25 per cent frequency is found from 200 samples, i.e. the observed number of quadrats containing the species is 50, the variance of the observed number is $200 \times \frac{1}{4} \times \frac{3}{4} = 37\cdot5$ and the standard error is $\sqrt{37\cdot5} = 6\cdot12$, so that the percentage frequency of 25 per cent has a standard error of 3·06. The binomial distribution,

37

unless n is large or p close to 0·5, is very asymmetric. For small values of n confidence limits cannot therefore be determined from the table of t. Mainland *et al.* (1956) give tables of the exact confidence limits at 95 per cent and 99 per cent probability levels for various values of n. A less extensive table is reproduced by Snedecor (1946, § 1.3). A small selection from the tables is given in TABLE 4 together with the values calculated by the use of t and the

TABLE 4

95% CONFIDENCE LIMITS FOR BINOMIAL DISTRIBUTION (AFTER SNEDECOR, 1946) AND LIMITS OBTAINED BY USING THE TABLE OF t (IN BRACKETS)

Percentage frequency observed	Number of observations				
	10	20	50	100	1,000
0	0 31	0 17	0 7	0 4	0 0
10	0 45	1 31	3 22	5 18	8 12
	(0 31)	(0 24)	(0 19)	(4 16)	(8 12)
20	3 56	6 44	10 34	13 29	18 23
	(0 49)	(1 39)	(6 31)	(12 28)	(18 22)
30	7 65	12 54	18 44	21 40	27 33
	(0 63)	(8 51)	(17 43)	(21 39)	(27 33)
40	12 74	19 64	27 55	30 50	37 43
	(5 75)	(17 63)	(20 54)	(30 50)	(37 43)
50	19 81	27 73	36 64	40 60	47 53
	(14 86)	(27 73)	(36 64)	(40 60)	(47 53)
60	26 88	36 81	45 73	50 70	57 63
	(25 95)	(37 83)	(46 74)	(50 70)	(57 63)
70	35 93	46 88	56 82	60 79	67 73
	(37 100)	(49 92)	(57 83)	(61 79)	(67 73)
80	44 97	56 94	66 90	71 87	77 82
	(51 100)	(61 99)	(69 84)	(72 88)	(78 82)
90	55 100	69 99	78 97	82 95	88 92
	(69 100)	(76 100)	(81 100)	(84 96)	(88 92)
100	69 100	83 100	93 100	96 100	100 100

standard error. Two features appear from this table, the inadvisability of using the standard error in assessing the accuracy of frequency values at the two extremes unless they are based on a very large number of throws, and the low level of precision of values derived from a relatively small number of samples. The latter point is often overlooked. It is not uncommon in the literature to find communities described in terms of percentage frequency values of different species obtained from twenty or even less examples. It will be seen from TABLE 4 that, for instance, even taking the not very extreme probability level of 95 per cent, an observed value of 50 per cent from 20 samples may arise from a true value anywhere between 27 and 73 per cent. It is clear that a sample number of at least 100 and preferably higher should be aimed at. If circumstances prevent this, it must be accepted that only gross differences between communities will be detectable. In view of the generally asymmetric form of the binomial distribution and the correlation between its variance and mean, it is not advisable to use a t test to compare two frequencies. Instead they may be compared by a contingency table. Suppose frequencies of 51 and 62 per cent respectively have been found from 200 samples in each of the two areas. The contingency table is

	Species present	Species absent	
Area A	a 102	b 98	200 $(a + b)$
Area B	c 124	d 76	200 $(c + d)$
	$(a + c)$ 226	$(b + d)$ 174	400 $(a + b + c + d = n)$

In this case the expected values, if the true frequencies are the same, are 113 in cells a and c, and 87 in cells b and d. The χ^2 testing departure from this expectation may be calculated in the normal way as the sum of $\dfrac{\text{deviation}^2}{\text{expected}}$ for each of the four cells. For such a 2 × 2 table it is, however, simpler to use the direct formula:

$$\chi^2 = \frac{(ad - bc)^2\, n}{(a + b)\,(c + d)\,(a + c)\,(b + d)}.$$

The resulting value is referred to the table of χ^2 with one degree of freedom. If n is small the numbers of possible tables with the same marginal totals is relatively small so that the distribution is discontinuous whereas χ^2 is a continuous distribution. The inaccuracy introduced by this may be allowed for by applying Yates's correction for continuity. For a 2 × 2 table this takes the form of subtracting 0·5 from each of the two values greater than expectation and adding 0·5 to those less than expectation. This correction should be applied if any expected value is less than 500 (Fisher and Yates, 1943). Thus, for the example quoted

$$\chi^2 = \frac{(76{\cdot}5 \times 102{\cdot}5 - 123{\cdot}5 \times 97{\cdot}5)^2\, 400}{226 \times 174 \times 200 \times 200} = 4{\cdot}486$$

which corresponds to a probability of between 0·05 and 0·02, indicating a significant difference at the conventional level of 5 per cent probability.

There is a further possible source of inaccuracy. If any marginal total is small (say less than 100), there is a marked difference in probability of a given deviation in the two directions. This asymmetry is allowed for in Fisher and Yates's (1943) Table VIII where 2·5 per cent and 0·5 per cent points for χ_c (square root of χ^2, corrected for continuity) are tabulated for the two tails of the distribution separately. 2·5 per cent probability for one tail, i.e. deviation in one direction, is equivalent to 5 per cent probability in the table of χ^2, where the probability given is for a particular deviation in either direction. For certain regions of the table no values can be given, owing to relatively large differences in probability for different marginal totals, and the exact solution (Fisher, 1941, § 21.02) must be used. The probability of getting one particular table with given marginal totals is, using the notation shown in the example above,

$$\frac{(a+b)!\ (c+d)!\ (a+c)!\ (b+d)!}{n!} \cdot \frac{1}{a!\ b!\ c!\ d!}$$

Thus it is possible to calculate exactly the probability of getting a difference as extreme, or more extreme than that observed if the two areas have in fact the same frequency. Consider two estimates of 92 per cent and 99 per cent frequency, each derived from 100 samples. Only one more extreme table, corresponding to frequencies of 91 per cent and 100 per cent, is possible and the required probability is

$$\frac{100!\ 100!\ 191!\ 9!}{200!} \left(\frac{1}{99!\ 8!\ 92!\ 1!} + \frac{1}{100!\ 9!\ 91!} \right) = 0·036$$

or 3·6 per cent, equivalent to 7·2 per cent in the table of χ^2. There is no indication that the two estimates represent different frequencies.

The methods of calculating the probability of observed differences have been outlined but in practice the probability can be obtained directly in nearly all cases from tables given by Mainland et al. (1956) and by Pearson and Hartley (1954).

If frequencies from several areas are to be compared to determine if they can reasonably be regarded as estimates of one value, a similar approach may be used. χ^2 is calculated from the contingency table, which will have degrees of freedom one less than the number of areas being compared. If comparisons of the means of sets of

40

estimates with one another are to be made, the data may be subjected to transformation and the ordinary analysis of variance procedure used. The appropriate transformation is to θ where $\sin \theta = \sqrt{}$ frequency expressed as a proportion. A short table of this angular transformation for steps of 1 per cent is given by Fisher and Yates (1943) and a fuller table by Bliss is reproduced by Snedecor (1946, § 16.7). The variance of θ is largely independent of p, depending mainly on the number of samples. The approximate value is $\dfrac{820\cdot7}{n}$ and this may be used if n is reasonably large (50 or over). If n is small this approximation deviates widely from the true value, particularly if p is very high or low. A table of variance for a selection of lower values of n is given in Appendix B (TABLE 1). Freeman and Tukey (1950) have suggested an arc-sine transformation, which is somewhat more effective in stabilizing the variance and may usefully be employed if n is relatively small. This takes the form

$$\theta = \tfrac{1}{2} \left\{ \arcsin \sqrt{\left(\frac{x}{n+1}\right)} + \arcsin \sqrt{\left(\frac{x+1}{n+1}\right)} \right\}$$

with variance tending to $\dfrac{820\cdot7}{n+\frac{1}{2}}$ (x = number of successes out of n trials). The transformation has been tabulated by Mosteller and Youtz (1961) for values of n up to 50.

As with density determinations, the accuracy of a frequency estimate can often be increased by suitable restriction of randomization. This can readily be seen by consideration of a hypothetical area, one half of which is completely blank and in the other half of which there are individuals so closely spaced that a quadrat of a certain size placed in it will always include at least one individual. The true value of the frequency with that size of quadrat will be 50 per cent. Sets of n random samples taken from it will show a range of value having a mean, within the limits of chance variation, of 50 per cent and falling on a binomial series $(\frac{1}{2} + \frac{1}{2})^n$. If the randomization is restricted in such a way that half the samples are taken randomly from the occupied half of the area and half from the unoccupied half (and this still gives every part of the area an equal chance of being represented) the value obtained will always be exactly 50 per cent.

If an area over which frequency is not uniform is divided up into a number of blocks of the same size, and if an equal number of quadrats is recorded in each block, a more accurate estimate of mean frequency over the whole area is likely to result than if the same number of samples were taken entirely randomly within the area. This procedure, which is little more laborious than complete randomization, is adequate if the frequency is being determined for purely descriptive purposes. It has the serious drawback that no estimate of the variance of the resulting frequency value is available, for the distribution is no longer strictly binomial. It cannot be greater than the corresponding figure for completely random samples. If it is desired to make exact comparisons with other areas, the procedure must be modified. If several sets, each of a smaller number of samples, are taken in the same way the variance of their mean may be calculated from the data. Since the data are of similar type to binomial distribution, angular transformation is still advisable, unless the number of occupied quadrats recorded in each set is large (say 100 or more). The treatment of the data may be illustrated by an example. Suppose five sets of 100 samples were taken, equal numbers of each set from the several blocks in the manner described, and showed 18, 18, 20, 21, 23 quadrats occupied respectively. The mean frequency is 20 per cent. The transformed values for the five sets are 25·1, 25·1, 26·6, 27·3, and 28·7 respectively. From these figures $Sx = 132·8$, $\bar{x} = 26·56$, $Sx^2 =$

$3536·56$, $S(x - \bar{x})^2 = 3536·56 - \dfrac{(132·8)^2}{5} = 9·392$, variance of

one set $\dfrac{9·392}{4}$, variance of mean of five sets $\dfrac{9·392}{4 \times 5} = 0·4696$,

standard error of mean $\sqrt{(0·4696)} = 0·685$. (For comparison, the variance of frequency from 500 random observations is, on the

transformed scale, $\dfrac{820·7}{500} = 1·64$, and the standard error 1·28.)

Note that the calculated standard error applies only to the mean on the transformed scale, and no standard error is available for the untransformed mean. Any desired confidence limits may, however, be calculated on the transformed scale and the corresponding frequency values obtained from the table of the angular transformation. Thus, for the example quoted the value of t for 5 per cent probability and four degrees of freedom is 2·78 which gives as the 95 per cent confidence limits for the transformed mean 26·56 $\pm$ (0·685 $\times$ 2·78) or 24·66 and 28·46, equivalent to percentage frequencies of 17·4 per cent and 22·7 per cent. For 500 random samples the corresponding limits are (from tables) 16·6 and 23·8 per cent.

Extensive use of completely systematic samples for frequency has been made by a group of Dutch workers concerned with artificial grassland (see De Vries and De Boer, 1959). Of particular interest is the demonstration, by Nielen and Dirven (1950), that the accuracy of such samples for grass fields is equivalent to that of random samples using the same number of quadrats.

The principles of sampling percentage cover are essentially the same as for frequency.* Indeed, in recording hits by point quadrats we are recording frequency sampled by quadrats of, theoretically, infinitesimally small size. Thus considerations of sample size, transformation of data before analysis and restriction of randomization are exactly the same as for frequency as long as points are positioned independently of one another. Three considerations only call for comment: (*a*) the effect of pin diameter; (*b*) the common practice of using frames of pins, often ten arranged in a line, instead of single pins; (*c*) the effect of inclining pins from the vertical. The effect of pin diameter has been mentioned in Chapter 1. The sample obtained by the use of a pin is not a point but a circular area of the same radius as the pin. Thus the true percentage cover is exaggerated, sometimes very considerably (Goodall, 1952b—cf. FIGURE 1 and TABLE 1). The error introduced is not the same for different species, as the percentage exaggeration will be greater for species having small or much elongated or dissected leaves than for species with large and more or less iso-diametric leaves. The presence or absence of gaps in the leaf mosaic of an individual will also influence the degree of exaggeration, an individual with few gaps between leaves tending to behave as a large single leaf. There is little prospect of any method being devised to allow in any exact manner for this varying exaggeration of cover between different species. Fortunately, in most circumstances it remains broadly constant for any one species and so will not affect comparisons between cover values for the same species in different communities. The same considerations apply more strongly to the 'loop-frequency' method of Parker (1950, 1951) in which presence in a ring three quarters of an inch in diameter is used as a measure of cover or basal area. Hutchings and Holmgren (1959) found for crown area of the shrub *Eurotia lanata* an exaggeration of up to 99 per cent compared with the value obtained by measurement of a sample of individuals. Johnston (1957) concluded that, for grassland, the loop-frequency method, while more rapid, gave less accurate results for comparable sampling intensity than either point-quadrats or line-intercept methods.

* The principles of the method have been thoroughly and critically discussed by Goodall (1952b).

Following Levy (Levy, 1933; Levy and Madden, 1933) most workers have used frames of ten pins. If cover were random, or aggregated only in such small units that the reading for one pin was independent of those for pins adjacent to it, this procedure would not seriously affect the accuracy of data thus obtained, though even then open to some theoretical objection because of the lack of statistical independence of the observations. In fact plants are commonly so arranged that there is pronounced interdependence between pins of the same frame. Thus, if the number of hits per frame for a series of frames are compared with binomial expectation, there tends to be an excess of high and low values and deficiency of values around the mean. This is to be expected from the clumped nature of much vegetation and it has been demonstrated by Goodall (1952b). It follows that the variance is greater than that from the same number of random samples and that the cover estimate is less precise. Goodall (*loc. cit.*) has shown for one community, not apparently with an unusual amount of non-randomness, that, in comparison with 2,000 points in random frames of ten points, from a quarter to a half this number of random single points, according to species, will give the same degree of precision. This consideration should be borne in mind in deciding whether to use frames or single points. In some circumstances it may be more economical of time to examine a greater number of points in frames than fewer points separately, but, in general, in view of the uncertainty before the observations are made of how much loss of precision will result from association of points into frames, separate points are to be preferred unless there is strong evidence against their use (e.g. from previous experience with similar vegetation). If frames are used variance must be calculated from the data, no prediction based on binomial theory being possible. The use of a completely systematic arrangement of points has been considered by Tidmarsh and Havanga (1955) in connection with the use of a 'wheel-point' apparatus in which one of the spokes of a rimless wheel is used as the sampling point for basal area. They showed, both for model populations of cards and in the field, that the variance of the mean obtained from systematic points tended to that for random points provided the spacing between points exceeded the size of individuals or clusters of individuals. (*Cf.* the similar conclusion of Nielen and Dirven, 1950, on systematic samples for frequency.) They therefore concluded that data gathered in this way, provided the spacing does not correspond with a repetitive pattern in the vegetation, could be treated as if they were obtained from the same number of random points. This has evident advantages in view of the generally greater speed of systematic sampling. They showed, further, that if the interpoint distance was less than the individual or cluster size, the variance between repeated independent systematic samples was less than that for random points. This greater accuracy is not spurious, as they claim, but is due to an effect comparable to that of restriction of randomization, discussed on p. 41. There is, however, no means of estimating the variance of a single set of systematic samples. The contrast may be noted with the use of random frames, where the variance is increased because the frame is normally small in relation to the individuals or clusters being recorded.

The use of inclined pins was suggested by Tinney *et al.* (1937) on the grounds that the larger sample of leaf contacts would lead to greater accuracy, and this suggestion has been fairly widely followed. It is true that the range of confidence limits for a given probability, expressed as a proportion of the cover value, decreases as cover increases (see TABLE 4), so that greater accuracy of the estimate is obtained from a higher cover. At the same time it must be remembered that the proportional increase in cover value obtained for different species by inclining the pins will vary according to the morphology of their shoots, as Winkworth (1955) has demonstrated in the field. Moreover, if shoots tend to be orientated in one direction, the cover value obtained may vary according to the orientation of the inclined pins. Thus, for the clearly understood, if somewhat arbitrary, character of vertical projection of the aerial parts, is substituted a much vaguer one. On balance there seems little advantage to be gained from inclining the pins, except perhaps when interest centres on species of very low cover.

If point quadrats are used to determine contribution of different species to the vegetation (proportion of total hits) as a measure of yield, or to determine cover repetition as a measure of performance, much the same considerations apply as in simple determination of cover. Single points will give a more accurate estimate than the same number of points in groups. Increasing the diameter of the pins will increase the values for cover repetition and may alter those for relative contribution to the vegetation, owing to the differential effect of pin diameter on species of different morphology. Cover repetition is likely to be increased for most species by inclining the pins, though this is not necessarily so and it may in fact be decreased (Winkworth, 1955). Winkworth has shown that, although mean values for percentage contribution to the vegetation in the heathland he examined do not differ significantly between determinations by vertical and inclined pins, the variance is liable to be greater. This is contrary to the apparent assumption of Tinney *et al.* (1937) that the increased number of contacts necessarily gives greater accuracy. Warren Wilson (1959a, b, 1960) has more recently made a critical examination of the effect of inclining point quadrats. He assumes that the objective in point quadrat techniques is always to determine the foliage area of a species per unit area of ground and is thus concerned primarily with cover repetition. He points out that vertical quadrats can in theory record any proportion of actual foliage area between 100 per cent (for horizontal foliage) and 0 per cent (for vertical foliage) but that an inclined quadrat never has such a wide range. He presents a formula giving the relation between apparent and true foliage area for different foliage angles and different angles of inclination of the quadrat. The angle of inclination giving an estimate of foliage area least affected by foliage angle is 21·5 degrees with the horizontal; with this inclination an estimate of foliage area per unit area of ground is obtained by multiplying the mean frequency of hits per quadrat by a factor of 1·1. He has further suggested analysing the foliage area at successive levels to bring out the relative amounts of foliage at different heights. To do this it is necessary to record the position at which a hit is made as the quadrat is moved into the vegetation. The average foliage angle (α) in any given layer can be obtained from records for vertical quadrats together with those for horizontal quadrats moving in the same layer, and is given by

$$\tan \alpha = \frac{\pi}{2} \left(\frac{\text{Contact frequency/cm of quadrat for horizontal quadrats}}{\text{Contact frequency/cm of quadrat for vertical quadrats}} \right).$$

A difficulty sometimes arises when the number of contacts per pin is recorded for tussock species. In the centre of a tussock the number of contacts may be too great to count. Goodall (1953c), arguing from the fact that where counts can be made at all points the data for number of points with one, two, three, etc. contacts can be fitted to a negative binomial series, has suggested using the negative binomial series calculated from the counted points to estimate the mean number of contacts at uncounted points. As he himself admits, fitting the negative binomial series to such data is laborious, and scarcely practical for routine work. In any case the assumption that the type of distribution in tussocking species is the same as in non-tussocking species is not necessarily true.

Measurement of cover or canopy by intersects on line transects calls for little comment. Randomization may conveniently be made by laying parallel transects from random points on an arbitrary base line or lines. An estimate of standard error is available from the variance of the total length of intercept on different transects. Greater accuracy will thus be obtained from more short transects than few long lines. Each transect should, however, be long enough to include all phases of any mosaic pattern that may be present. McIntyre (1953) has considered the possibility of making a concurrent estimate of density (especially of tussock or similar growth forms) from the number of separate intercepts on the transects. He concludes that, in most circumstances, there is little to be gained over making separate counts for density determination.

Comparison of estimates of cover, if they are based on random point quadrats, may be made by a 2×2 contingency table in the same way as comparison of frequency estimates. If frames of pins have been used, the assumption of binomial distribution is no longer valid and this procedure is liable to exaggerate the significance of differences. Instead the values from individual frames should be subjected to angular transformation and the variance of the means for areas calculated and used as the basis of a t test. Data for percentage contribution to the vegetation may be handled in the same way. Data for cover repetition present a different problem. Here the distribution is very skew. Goodall (1952b) has found empirically that square root transformation will put such data into a form satisfactory for the standard tests of significance.

In sampling performance by measures such as height, cover repetition, etc., which refer to the individual plant, samples should ideally be taken in such a way that all plants have an equal chance of being selected. Unless density is uniform over the area, this can only be achieved by enumerating the individuals and selecting individuals by use of random numbers. This is clearly not practicable in most circumstances. If density is reasonably uniform over the area, the best approximation is to select the individuals nearest to a number of random points. If density is obviously variable over the area, it is advisable to divide the area into portions each having roughly constant density, determine separate means for each portion, and combine the values in a weighted mean, each value being weighted by the number of individuals present in the sub-area, i.e. by the product of density and area. The effect of this may be illustrated by an example. Suppose counts are made of number of inflorescences per individual, and the number, as is likely, varies inversely with the density of individuals, results such as the following might be obtained in different parts of the area.

47

	A	B	C	D
Relative area	1	2	1	4
Density	5	12	18	25
Counts of inflor-escences per individual	3, 4, 2, 2, 5 1, 4, 3, 2, 4	5, 3, 1, 3, 4 4, 3, 2, 2, 1	2, 4, 3, 2, 3 1, 3, 3, 1, 2	1, 3, 2, 1, 1 2, 4, 2, 1, 3
Mean numbers of inflorescences per individual	3·0	2·8	2·4	2·0
Weighting (relative area × density)	5	24	18	100

The weighted mean will then be
$$\frac{3\cdot0 \times 5 + 2\cdot8 \times 24 + 2\cdot4 \times 18 + 2\cdot0 \times 100}{5 + 24 + 18 + 100} = \frac{325\cdot4}{147} = 2\cdot21.$$
The arithmetic mean of the 40 observations is
$$\frac{30 + 28 + 24 + 20}{40} = 2\cdot55,$$
which gives a quite misleading value for the average number of inflorescences, being unduly influenced by the relatively few plants with a large number of inflorescences.

There is a useful approximate test, derived from the rank of separate observations, of the significance of differences between means. This may be illustrated from an example. Ten frames of ten pins gave the following readings for cover of a species in two different areas

A. 10, 9, 9, 8, 7, 6, 5, 4, 4, 2. *B.* 7, 6, 5, 4, 4, 3, 3, 2, 1, 1.
Is there evidence of a difference in percentage cover? The values observed are rearranged in descending order

Rank   1 2 3 4 5 6 7 8 9 10 11 12 13 14 15 16 17 18 19 20
Observed   —   — — — - — — — — — —
value   10 9 9 8 7 7 6 6 5 5   4 4 4 4   3 3   2 2   1 1

The total rankings for the two areas are

A. $\quad 1 + 2 \times 2\cdot5 + 4 + 5\cdot5 + 7\cdot5 + 9\cdot5 + 2 \times 12\cdot5 + 17\cdot5 = 75.$

B. $\quad 5\cdot5 + 7\cdot5 + 9\cdot5 + 2 \times 12\cdot5 + 2 \times 15\cdot5 + 17\cdot5 + 2 \times 19\cdot5$
$\quad\quad = 135.$

Where there is a tie between two or more ranks each value is allotted the mean ranking. If there is no difference in mean cover, the expected total rankings for the two areas are the same, 105. The 5 per cent and 1 per cent points for the lower total are approximately $\dfrac{9\mathcal{N}^2}{10} - \dfrac{3\mathcal{N}}{2} + 3$ and $\dfrac{4\mathcal{N}^2}{5} - 9$, where $\mathcal{N}$ is the number of values in each set.* In this case the 5 per cent point is 78 and the 1 per cent point 71, so that significance at the 5 per cent level, but not at the 1 per cent level, is indicated. The t test, carried out after angular transformation, and entering the table at 9 degrees of freedom to allow for the markedly different variances of the two sets, indicates a probability of approximately 2 per cent.

This ranking test is quickly performed and not only provides a rough test in the field and an indication whether more laborious tests are worth applying, but it has the additional advantage that it is independent of the distribution of the variables being compared. It can thus be used even where it is impossible to transform the data into approximately normal form and the more usual tests are therefore not available.

Considerable attention has recently been paid to a method of plotless sampling, particularly adapted to forest vegetation, where there are practical difficulties in delimiting the relatively large quadrats necessary for sampling trees (Cottam, 1947; Cottam and Curtis, 1949, 1955, 1956; Cottam *et al.*, 1953). From a number of randomly selected points certain measurements are made. Four different procedures have been used (FIGURE 10).

1. *Closest individual method.* The distance from the sampling point to the nearest individual is measured (FIGURE 10(a)).

2. *Nearest neighbour method.* The distance from this individual to its nearest neighbour is measured (FIGURE 10(b)).

Theoretically (Morisita, 1954; Clark and Evans, 1954a) the mean value obtained by either method is half the square root of the mean area. Cottam *et al.* (1953) have confirmed this relationship empirically for the first method, but Cottam and Curtis (1956) found that the correction factor for the second method should be 1·67 instead

---

* Dixon and Massey (1957) have a useful discussion of this and other ranking tests and give a table of the probability of various departures from expected total rankings. The approximate 5 per cent and 1 per cent points are quoted from Moroney (1951).

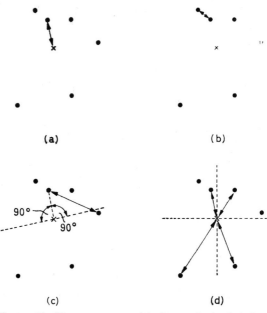

FIGURE 10. Distances measured in four methods of plotless sampling. (a) Closest individual. (b) Nearest neighbour. (c) Random pairs with 180° exclusion angle. (d) Point-centred quarter. X is the sampling point in each case

of the theoretical 2. The discrepancy arises because the method does not provide a random sample of nearest neighbour distances. The correction factor was determined empirically from a synthetic random population and should not be used uncritically. As Pielou (1959) has pointed out, it is unlikely that a single correction factor can be found which will always allow for the bias in the sample of distances.

3. *Random pairs method.* From the sampling point a line is taken to the nearest individual and a 90° exclusion angle erected on either side of it. The distance from this individual to the nearest one lying outside the exclusion angle is measured (FIGURE 10(c)). The correction factor to obtain the square root of the mean area is 0·8 (Cottam and Curtis, 1955).

4. *Point-centred quarter method.* The distance from the sampling point to the nearest individual in each quadrant is measured. The orientation of the quadrants is fixed in advance. Cottam and Curtis (1956), sampling at random distances along a transect line, orientate them two on either side of the line (FIGURE 10(d)). The

mean of all distances measured has theoretically (Morisita, 1954) and empirically (Cottam *et al.*, 1953) been shown to be equal to the square root of the mean area.

From the mean value of the measurements made by any of these methods, the mean area, and thus the density, of all individuals may be calculated. There is one proviso, viz. the methods all assume that individuals are randomly distributed. Comparison by Cottam and Curtis (1956) with complete enumerations of woodlands indicates that this assumption is justified, at least for the areas examined. It would not necessarily be so in all cases and almost certainly not if the methods were applied to a single species in mixed woodland. Dix (1961) has used the point-centred quarter method in grassland, with individual shoots as the unit of density. To do so may be valid in some grasslands but, in view of the evident non-randomness of many grassland species, the method should be tested against direct counts for any grassland type to which it is to be applied.

Morisita (1957) has suggested a method, applicable to certain types of non-random distribution, of estimating the density. The distance r to the nth nearest individual ($n \equiv 3$) in each of k sectors at N points is measured. He derives two estimates of density.

$$d_1 = \frac{1}{\pi} \cdot \frac{n-1}{N} \sum \left(\frac{1}{r^2}\right)$$

$$d_2 = \frac{1}{\pi} \cdot \frac{nk-1}{N} \sum_{i=1}^{N} \frac{k}{\sum^k_{j=1} r_{ij}^2}$$

If $d_1 < d_2$ the best estimate of density is $(d_1 + d_2)/2$. If $d_1 > d_2$, d_1 is the best estimate.

This approach is based on the assumption that the whole area sampled can be regarded as made up of a number of sub-areas within which individuals are randomly or uniformly distributed. It is thus likely to be applicable to a mosaic of patches of different densities but not to a distribution of clumps each composed of relatively few individuals. Morisita has tested the method with $n = 3$ and $k = 4$ on artificial populations and obtained satisfactory estimates of density. No field test has apparently been made.

In addition to measurement of distances the species and basal area of each tree involved are recorded. From these data figures for relative density and relative basal area of each species are readily obtained. It must be emphasized that these are valid whether the distribution is random or not (though the accuracy with which they are determined may, except in the first method, be affected by departure from randomness). Only if they are multiplied by the estimate of total density to obtain absolute values, may errors due to non-random distribution enter. Since, in many contexts, relative is of greater interest than absolute comparison, the method is clearly valuable even if only the species present and basal area are recorded for each sample point and no measurements of distance made.

Cottam and Curtis derive from the last three methods a measure of relative frequency (frequency recorded in samples as a proportion of frequency of all species). Like quadrat frequency, this is dependent partly on density and partly on type of distribution of the species concerned. It is, however, further affected by the density and pattern of other species present, and thus becomes an even more complex character of the vegetation and correspondingly difficult to interpret. It is true that relative density is similarly dependent on other species but there is the important difference that density is an absolute character, unlike frequency, and the corresponding absolute value can be obtained by multiplying by the estimated density.

Cottam and Curtis (1956) show that the accuracy of the estimate of density, for the same number of points sampled, increases in the order: closest individual, nearest neighbour, random pairs and quarter method. The quarter method involves making a greater number of measurements to obtain a given accuracy than the random pairs method, but at considerably fewer points. They suggest its use, however, in preference to random pairs as occupying less time in the field and giving a greater amount of information on relative density and basal area.

No use has apparently yet been made of these methods in exact comparisons. Estimates of total density cannot readily be compared but instead the actual measurements made may be used as a measure of mean area, their means being compared by a *t* test or analysis of variance. Cottam *et al.* (1953) show that the mean value of the four measurements at a point in the quarter method is normally distributed and hence this measure may be analysed directly. If random pairs are used square root transformation is advisable to reduce skewness. Relative density in different areas may be examined in a contingency table (cf. percentage contribution to the vegetation from point quadrats). Relative frequency, if it is thought useful, could be handled in the same way. The data for basal area are less tractable. Comparison of mean basal area per tree, either for total trees or for a species, are readily compared by a *t* test, but there is no ready means of comparing total basal area of trees per unit area, which involves the multiple of two values of which estimates only are available. Shanks's (1954) calculation of estimates of density and mean basal area per tree from each point scarcely commends itself, as both estimates rest on quite inadequate samples and the resulting high standard error could only be reduced by using such large samples that the advantage of speed in the field would be lost.

If basal area is the measure of greatest interest, the method of 'variable-radius' sampling introduced by Bitterlick (1948; see Grosenbaugh, 1952a,b, Shanks, 1954, and Palley and Horwitz, 1961) is more useful. In this method a sighting gauge is used to count those trees which are within a distance from the sampling point not more than 33 times their trunk diameter, i.e. which subtend an angle of at least 1° 44′ at the point. This ratio is so fixed that the count at a point multiplied by 10 corresponds directly to square feet of basal area per acre. The values obtained are approximately normally distributed and hence can be analysed directly. It must be emphasized that no other measures of any value are available from variable-radius sampling. The relative frequency figures, calculated by Rice and Penfound (1955), represent such a complex character, affected not only by density and pattern of individuals of the species concerned and of those of other species present, but also by the pattern of size classes of the different species, as to be virtually meaningless.

Lindsey et al. (1958) have made a careful assessment in forest of the efficiency of various areal and plotless sampling methods for both density and basal area. Their comparison is based on determination of the time necessary to sample sufficient units to reduce the standard error of the resultant estimate to an arbitrary level of 15 per cent. They took into account not only time at sampling points but also time spent in moving between them. The last is clearly an important consideration but is often overlooked in discussions of efficiency of different methods. Their conclusion is that Bitterlick variable-radius sampling is to be preferred for basal area and that a 0·1 acre circular plot defined by rangefinder is to be preferred for density.

53

CHAPTER 3

PATTERN

ECOLOGY and plant geography are largely concerned with the causes of patterns of distributions, patterns of all scales from those of individuals within a small area to those of vegetation types and taxa over the surface of the world. One of the principal contributions that may be expected from the use of quantitative methods in ecology is the more exact detection and description of distribution patterns. Large-scale patterns such as those which interest the plant geographer are readily recognized qualitatively. Quantitative methods are, therefore, not required, though they may in time prove a useful tool in comparison of differing patterns. Such problems are in any case the concern of the plant geographer rather than the ecologist and are outside the scope of this book.

The first work on non-randomness was naturally exploratory and able to contribute relatively little to ecological knowledge, but techniques of analysis of non-randomness now available are more useful. Before discussing technique the biological significance of departure from randomness must be examined.

A living plant and its various parts are subject to a multitude of variable factors, all affecting to greater or lesser extent its behaviour and performance, and even its continued existence. These factors include internal ones intrinsic in the plant and environmental ones. The external factors range from such obvious and measurable influences as temperature, humidity, and concentration of nutrients, through obvious but less readily described features such as soil texture to such little-understood effects as specific secretion from other plants. Intrinsic factors fall principally under the heads of effect of position of the part on the whole plant, and effect of age. Effective seed dispersal distance is also best considered an intrinsic factor. The effects of the numerous influencing factors are not independent but exhibit numerous and complex interactions of which as yet we know comparatively little. Some few are well known as phenomena, though understanding of their mechanism is only beginning, e.g. the effect of concentration of one nutrient on the availability of another. Others have been singled out by plant

54

physiologists for special study, e.g. the interaction of light, temperature, and carbon dioxide supply in photosynthesis. However, we lack more than the most superficial knowledge of many of the factors affecting the performance of plants in the field.

In any given area, of whatever size, some factors will be constant at any one time, while others will vary from point to point in the area. The smaller the area the greater in general will be the number of constant factors. The magnitudes of the ranges of difference in the varying factors may be considered in terms of their effects on the plant. If the effects of all factors on all species present are relatively small, it will be a matter of chance which species succeeds at any point, and the resulting distribution of individuals (or parts of individuals where such parts are largely independent) will be random. Such conditions of equality of effects of different factors will only hold when the range of values found is well within the limits of tolerance of all the species present. This follows from the much greater effect of small differences in an influencing factor when its value is near the limits of tolerance of a species. Now if one or few factors have a disproportionately great effect on performance or survival of a species, then the distribution of that species will tend to be determined by that particular factor or factors. If the values of the factor are themselves randomly distributed then the distribution of the species will also be random. Field experience shows that most environmental factors do not have a random distribution of different values (consider, for example, such factors as soil moisture and texture). We may thus put forward the hypothesis that departure from randomness of distribution of a species indicates that one or few factors are determining the performance or survival of the species. The converse does not necessarily apply. One factor may be overriding even if the species is randomly distributed. It is unlikely, however, in any community of more than a few species, that the factor will have a predominant influence on one species alone. If more than one species is concerned, whether the effect of the predominating factor is the same for different species or not, correlation between the occurrence or performance of species may be expected. (See Chapter 4.) Thus correlation gives an indication of predominance of one or few factors, even if individuals are randomly distributed.

Since study of the causal factors determining the distribution of plants and vegetation is a prime objective of ecology, any technique that can assist their detection is clearly of value. At the same time it must be emphasized that detection and analysis of non-randomness is a starting-point for further investigation of the factors responsible and not an end in itself. Failure to realize this is responsible for the apparently barren nature of some statistical work in ecology.

Though the pattern of distribution of a species within a community is a real characteristic of that community, the appearance of non-randomness in a set of samples is not an absolute characteristic but, like frequency, is dependent on the size and, sometimes, the shape of sample area used. This may be illustrated from FIGURE 11,

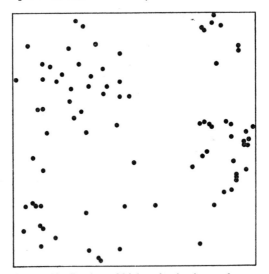

FIGURE 11. Patches of higher density imposed on a general distribution at lower density, a type of distribution commonly found in the field

which shows a distribution of individuals with a general low density but scattered patches in which the species has a high density. A relatively small quadrat will clearly tend to be unoccupied more often in such a community than in one with the same density of individuals randomly distributed. Correspondingly, high values will occur more frequently also. There will, in fact, be clear evidence of non-random distribution. With a very small quadrat, however, so small that even when placed within the high-density areas it is usually unoccupied, the position is rather different. The absolute increase in number of unoccupied quadrats and of those containing more than one individual will be so small that it will not be detected unless a very large number of samples is taken. Thus the non-randomness will not be apparent. If a large quadrat is used, such that it generally includes one or several high-density patches, the grouping together of individuals will tend to affect only their distribution within the quadrat and not the number it contains. The distribution, judged by such samples, will thus tend to appear random.* These conclusions can readily be checked on artificial communities of counters of known distribution (cf. Greig-Smith, 1952a). Many types of non-random arrangement sampled by random quadrat throws will thus appear random with very small or large quadrats but non-random with intermediate sizes. It is also clear that the lower the density of a species within high-density patches, the greater is the minimum size or number of quadrats necessary to detect departure from randomness. It is commonly stated that the rarer species in a community, unless propagated vegetatively, are randomly distributed, whereas the commoner species are usually not randomly distributed. This conclusion may result in part from this relationship between density and minimum quadrat size and number necessary to detect non-randomness and should not be accepted uncritically.

The distribution of individuals may be more regular than random expectation though this is very much less commonly found. If they are so distributed, very small quadrats will again fail to detect departure from randomness, but all larger quadrat sizes will do so and, in general, the larger the quadrat the more distinct the departure from random expectation will appear.

So far we have not questioned the appropriateness of the Poisson series as the basis of random expectation. The condition under which a Poisson distribution applies is that the chance of an event occurring is very small but if a sufficiently large sample is taken then some occurrences will be found. For quadrat data the relevant chance is that of a plant occurring at any point in the area under consideration. Provided this is low then a Poisson distribution may be expected if the area is sampled by any size of quadrat small relative to the whole area. In biological terms the expectation applies if the number of individuals of a species in the area is low relative to the possible number that could grow in that area. It should be noted carefully that the mean number of individuals per quadrat is irrelevant in deciding whether a Poisson expectation is appropriate. If the mean number per quadrat is, say, 10 or a hundred, Poisson expectation still applies if the maximum possible number that could occur in a quadrat is very much higher.

If the number of individuals actually occurring approaches the maximum possible, the Poisson expectation no longer applies.

* Whether the data from such a large quadrat fit random expectation exactly will depend on the pattern of the high-density patches (cf. Goodall, 1952a, p. 205).

Instead, if individuals are randomly distributed, the frequencies of different numbers per quadrat will approximate to a binomial distribution obtained by expansion of $(p + q)^n$, where n is the maximum number possible per quadrat, p the chance of any one of the possible 'places' in the quadrat being occupied (so that np is the mean number per quadrat) and $q = 1 - p$. The binomial series has $n + 1$ terms, giving the probability of $0, 1, 2, \ldots n$ individuals per quadrat, e.g. if the quadrat can contain a maximum of four individuals the expansion is

$$q^4 + 4pq^3 + 6p^2q^2 + 4p^3q + p^4.$$

TABLE 5

PROBABILITIES OF FINDING 0, 1, 2, 3, ETC. INDIVIDUALS IN
SAMPLES FROM VARIOUS DISTRIBUTIONS WITH MEAN OF 3

Number of individuals per sample	Distribution			
	Poisson	Binomial $(\frac{1}{3} + \frac{2}{3})^9$	Binomial $(\frac{1}{2} + \frac{1}{2})^6$	Binomial $(\frac{3}{4} + \frac{1}{4})^4$
0	0·0498	0·0260	0·0156	0·0039
1	0·1494	0·1171	0·0937	0·0469
2	0·2240	0·2341	0·2344	0·2109
3	0·2240	0·2731	0·3125	0·4219
4	0·1680	0·2048	0·2344	0·3164
5	0·1008	0·1024	0·0937	
6	0·0504	0·0341	0·0156	
7	0·0216	0·0073		
8	0·0081	0·0009		
9	0·0027	0·0001		
>9	0·0012			
Variance	3·00	2·00	1·50	0·75

TABLE 5 shows for a mean of 3 the probabilities of different numbers per quadrat given by the Poisson series and by binomial series corresponding to $\frac{1}{3}$, $\frac{1}{2}$, and $\frac{3}{4}$ of the maximum possible number of individuals. It will be seen that at these levels of relative density there is considerable discrepancy between the binomial and Poisson series, the discrepancy increasing as the density approaches the maximum possible. In each of the binomial series there is a greater proportion of values at and around the mean and fewer extreme values than in the Poisson series.

The density relative to maximum possible and, correspondingly, the maximum number of individuals that could occur in a quadrat of any specified size are theoretical concepts incapable of determination, though of some interest. A rough estimate may be possible from a knowledge of the morphology and behaviour of a species.

What is important in the present connection is some indication of the relative density at which the Poisson series no longer gives a reasonable approximation to random expectation. Fortunately this may be easily obtained.

The variance of a Poisson series is equal to its mean. A measure of the degree of departure from Poisson expectation is therefore provided by the ratio of the variance to the mean. The departure of this ratio from unity has a standard error $\sqrt{\dfrac{2}{N-1}}$, where N is the number of observations, and may be tested for significance by a t test.

Now a binomial series $(p + q)^n$ has a variance npq, which is always less than the mean np. The variance : mean ratio is thus

$$\frac{npq}{np} = q = 1 - p.$$

If this value is to be tested for departure from unity (on the assumption that the distribution is in fact Poissonian) we can put

$$1 - p = 1 - ts$$

and thus $p = ts$ gives the value of p, or proportion of maximum possible density at which difference between binomial and Poisson expectation would be just detectable at any desired level of probability. $S \left(= \sqrt{\dfrac{2}{N-1}} \right)$ is the standard error of the difference between the variance : mean ratio and unity, and t has the value (from table of t) corresponding to the number of observations and the desired level of probability.

$$
\begin{array}{lll}
\text{For 100 observations} & s = 0\cdot1421 \\
\text{500 observations} & s = 0\cdot06331 \\
\text{1,000 observations} & s = 0\cdot04474 \\
\text{For 5 per cent probability} & t = 2\cdot0 \\
1 \quad\text{,,} \quad\text{,,} & t = 2\cdot6 \\
0\cdot1 \quad\text{,,} \quad\text{,,} & t = 3\cdot4 \text{ for } N = 100 \\
& 3\cdot3 \text{ for } N = 500 \text{ or } 1,000
\end{array}
$$

Using these values we obtain the following approximate values of p:

N	Probability (per cent)		
	5	1	0·1
100	0·28	0·37	0·48
500	0·13	0·16	0·21
1,000	0·09	0·12	0·15

If 5 per cent probability is accepted as indicating significance, then with 100 throws significant departure from Poisson expectation would be expected to be detected in 50 per cent of trials in a random population if density was equal to 28 per cent of maximum possible or higher, but if 500 throws were used this would apply if density was only 13 per cent. If the more extreme probability level of 1 per cent was required before departure was accepted, density could be as high as 37 per cent for 100 throws, or 16 per cent for 500 throws, without departure being detected in 50 per cent of trials. It will be noticed that this relationship between binomial and Poisson expectations is independent of quadrat size.

The expected distribution at densities approaching the maximum may be considered in another way. Since an individual has finite size there is always an area around its midpoint (at which it is conventionally regarded as situated) within which the chance of another individual occurring is depressed because two individuals cannot be superimposed. If the density relative to maximum possible is low, the areas within which probability of occurrence is so reduced are small relative to the whole area and do not affect significantly the even distribution of probability over the area. If relative density is high this is no longer true; the distribution of probability of occurrence becomes markedly uneven and Poisson distribution no longer occurs.

If individuals vary widely in size the position is more complicated. At maximum density the relatively small spaces between large individuals may be occupied by relatively large numbers of small individuals, giving a pattern of individuals more uneven than the corresponding random distribution (Pielou, 1960). Though this might, in theory, complicate interpretation of non-random pattern, in practice it is unlikely to be important. It probably occurs in natural vegetation only rarely, principally in freely regenerating forest composed entirely or almost entirely of one species. Moreover, the practising ecologist is unlikely to treat individuals of such a wide range of sizes on the same basis.

There are two categories of departure from randomness, (1) that in which individuals tend to be clumped or aggregated together (FIGURES 2(c) and 11), sampling of which gives an excess of blanks and high values compared with random expectation, and (2) that in which individuals tend to be uniformly spaced (FIGURE 2(a)), sampling of which gives a deficiency of blanks and high values and an excess of values around the mean. The former has been termed *overdispersed* (referring to the distribution curve obtained) and the latter *underdispersed*. Unfortunately these two terms have sometimes been used in the reverse sense (referring to the pattern of individuals on the ground). They have thus become a source of confusion and are better avoided. Recently *contagious* has been widely used for aggregated distributions generally, and this seems a suitable term, although originally applied by Pólya to a particular type of aggregation. The uniform distributions may be termed *regular*. Regular distributions have rarely been found in the field and attention has rightly been concentrated on contagious distributions.

Various tests and measures of departure from randomness have been proposed and the more important are detailed below. To demonstrate their use it will be convenient to consider their application to one set of data. The following results were obtained for number of shoots of *Carex flacca* per quadrat in 200 throws of a quadrat 10 cm square in limestone grassland.

Number of shoots per quadrat	0	1	2	3	4	5	6	7	8	9	10
Number of quadrats	134	34	12	8	8	—	1	1	1	—	1
Mean number per quadrat	0·725										

(1) χ^2 *test of goodness of fit.* The Poisson series with the observed mean is calculated and the observed numbers of quadrats containing 0, 1, 2, etc., individuals compared with random expectation by a χ^2 test. This has been used by Blackman (1935) and many later workers.

For the *Carex flacca* data $m = 0.725$ and $e^{-m} = 0.48433$. (See Appendix B, TABLE 3.) Thus the expected number of empty quadrats is $200 \times 0.48433 = 96.866$. The remaining terms of the Poisson series are readily calculated in the manner shown on page 62.

It is necessary to lump together all the figures for quadrats containing three individuals or more, so that all the expected values may be greater than 5. (This arbitrary level is that commonly used in calculations of χ^2 to eliminate the disproportionate effect of small differences from a low expectation.*) The total value is entered in the table of χ^2 with degrees of freedom 2 less than the number of values from which χ^2 is calculated, 1 degree of freedom being used in the determination of the mean. In this case there are 2 degrees of freedom available for the χ^2 test and the 0·1 per cent point is 13·81. The probability of obtaining the observed distribution by chance is thus much less than 0·1 per cent and the data show very clear indication of non-random distribution.

(2) *Variance : mean ratio* [Coefficient of dispersion (Blackman, 1942)

---

* Cochran (1954), in a useful general discussion of χ^2, considers the generally recommended minimum expected value of 5 to be too conservative, resulting in a substantial loss of power in the test.

Number per quadrat	Number of quadrats		Difference	χ^2
	Expected	Observed		
0	$200e^{-m} =$ $200 \times 0 \cdot 48433 = 96 \cdot 866$	134	37·134	14·24
1	$200me^{-m} =$ $96 \cdot 866 \times 0 \cdot 725 = 70 \cdot 228$	34	36·228	18·69
2	$\dfrac{200m^2e^{-m}}{2!} =$ $70 \cdot 228 \times \dfrac{0 \cdot 725}{2} = 25 \cdot 458$	12	13·458	7·11
3 >3	$\dfrac{200m^3e^{-m}}{3!} =$ $25 \cdot 458 \times \dfrac{0 \cdot 725}{3} = \left.\begin{matrix} 6 \cdot 152 \\ 1 \cdot 296 \end{matrix}\right\} 7 \cdot 448$	20	12·552	21·15
	200	200		61·19

or relative variance (Clapham, 1936)]. This test makes use of the equality of mean and variance of the Poisson distribution. If the ratio of variance to mean is less than one, a regular distribution is indicated, if greater than one, a contagious distribution.

Two ways of testing the significance of the difference between the observed ratio and unity are available. The difference may be compared with its standard error by means of a t test. The standard error is independent of the density of individuals and depends only on the number of samples. Blackman (1942) used

$$s = \sqrt{\left[\frac{2N}{(N-1)^2}\right]},$$

where N is the number of samples. Bartlett (quoted in Greig-Smith, 1952a) has pointed out that the more correct value is $s = \sqrt{\left(\dfrac{2}{N-1}\right)}$. The difference is small for reasonably large values of N. A table of $\sqrt{\left(\dfrac{2}{N-1}\right)}$ for selected values of N is given in Appendix B (TABLE 4).

An alternative test of significance makes use of the index of dispersion $\dfrac{S\,(x-\bar{x})^2}{\bar{x}}$ i.e. $\dfrac{\text{variance}}{\text{mean}} \times (\text{number of observations} - 1)$.

Significance is assessed by reference to the table of χ^2, the observed value of the index of dispersion being entered in the table with degrees of freedom one less than the number of observations. This method of testing the variance : mean ratio is preferred by Skellam (1952) and David and Moore (1954).

For the *Carex flacca* data

$$Sx = 145,\; Sx^2 = 531,\; S\,(x-\bar{x})^2 = 531 - \frac{(145)^2}{200} = 425\cdot875$$

$$\text{Variance } \frac{425\cdot875}{199} = 2\cdot140075,\quad \frac{\text{Variance}}{\text{Mean}} = 2\cdot9518.$$

The standard error of the deviation of this value from unity is

$$\sqrt{\left(\frac{2}{199}\right)} = 0\cdot1003.$$

Thus $t = \dfrac{2\cdot9518 - 1}{0\cdot1003} = 19\cdot46$, with 199 degrees of freedom and probability (from table of t) much less than 0·1 per cent.

The index of dispersion is $I = 2\cdot9518 \times 199 = 587\cdot41$ with 199 degrees of freedom. The usual table of χ^2 includes values up to 30 degrees of freedom only. For higher degrees of freedom it may be assumed that $\sqrt{(2\chi^2)}$ is normally distributed about $\sqrt{(2N-1)}$ with unit variance, so that the value of $\sqrt{(2\chi^2)} - \sqrt{(2N-1)}$ may be referred to the table of the normal deviate (Fisher, 1941 § 20). The probability for χ^2 corresponds to that for a single tail of the normal curve so that the probability obtained from the table must be halved. Thus for the *Carex flacca* data

$$\sqrt{(2\chi^2)} - \sqrt{(2N-1)} = \sqrt{1,174\cdot82} - \sqrt{399} = 14\cdot301.$$

From the table of the normal variate the corresponding probability is very low (χ^2 probability of 0·05 per cent corresponds to a value of 3·29).

If deviation from random expectation is towards a regular distribution, the value of the index of dispersion is judged significant if the probability from the χ^2 table is unusually high, e.g. probability of 95 per cent corresponds to 5 per cent for contagious distribution and 99 per cent to 1 per cent. If the procedure for large number of degrees of freedom is followed, the value of $\sqrt{(2N-1)} - \sqrt{(2\chi^2)}$ is used and a low probability required for significance.

Numata's coefficient of homogeneity (Numata 1949, 1954), which has been used in various investigations by Numata and others, is related to the variance : mean ratio. It is defined for a sample of n quadrats with density $\bar{x}$ and standard error s as

$$h = t \cdot \frac{1}{\sqrt{n}} \cdot \frac{s}{\bar{x}},$$

t being assigned the value corresponding to a desired probability. The value obtained clearly depends on the density and number of samples as well as on the departure from randomness. When expressed, as Numata suggests, as the ratio of the observed value to the value expected for a random distribution of the same density sampled by the same number of quadrats it reduces to $\sqrt{}$(variance: mean ratio).

(3) *Moore's ϕ test* (Moore, 1953). If the mean number of individuals per quadrat is high, or if the distribution is markedly contagious, some quadrats will contain a large number of individuals which are both time-consuming to count and liable to counting errors. Moore has proposed a test dependent on the first three frequency classes (0, 1, 2 individuals per quadrat) only. He takes

$$\phi = \frac{2n_0\,n_2}{n^2_1}$$

where n_0, n_1, n_2 are the numbers of quadrats containing 0, 1, 2 individuals respectively. For a Poisson distribution $\phi = 1$. For contagious distributions with an excess of empty quadrats and deficiency of quadrats containing one individual ϕ will generally, depending on the value of n_2, exceed unity. At least we can infer that if ϕ is significantly greater than unity, then the distribution is non-random. (Some regular distributions will also give $\phi > 1$.) The mean value of ϕ for samples from a Poisson distribution, and its standard error, which is dependent on the mean and number of quadrats, have been derived by Moore. He gives a table (Appendix B, TABLE 5) showing the value of ϕ plus twice its standard error, i.e. approximately the 5 per cent point, for various values of N and the mean, m. Determination of m depends on complete enumeration of the quadrats. However, for each value of m there is a unique value of R, the percentage of quadrats falling in the first three classes, and the table may be entered according to the value of R instead of that of m. This is equivalent to estimating the mean from the first three classes only.

For the *Carex flacca* data $n_0 = 134$, $n_1 = 34$, $n_2 = 12$, $\phi = \dfrac{2 \times 134 \times 12}{34^2} = 2 \cdot 782$, $R = \dfrac{134 + 34 + 12}{200} \times 100 = 90$. The 5 per cent significance point for ϕ, from the table, is approximately

1·66 (nearest tabulated points $R = 92$, $\phi = 1\cdot66$ and $R = 81$, $\phi = 1\cdot68$) and hence non-random distribution is indicated.

(4) *Ashby and Stevens's test* (Ashby, 1935). This depends on random sampling by a quadrat subdivided into a number of smaller squares and comparison of the observed number of empty squares with the number expected from the density within the quadrat, which is

$E = n\left(1 - \dfrac{1}{n}\right)^s$, where n is the number of squares and s is the

total number of individuals in the quadrat. Although a comparison between observed and expected number of empty squares is available from a single quadrat sample, the distribution of the number of empty squares is so far from normal that data from a number of quadrats must be pooled before making a test of significance. The reader is referred to the original account for details of the test of significance.

(5) David and Moore (1957) have suggested a test based on division of an area into parallel strips and comparison of between strip and within strip variance of distance of individuals from the end. As this depends on knowledge of the exact position of all individuals in terms of co-ordinates from two adjacent sides of the area taken as axes, its ecological utility is limited.

(6) Aberdeen (1958) has pointed out that if frequency data are available from several sizes of quadrat, any departure from linearity in a graph of log percentage absence against quadrat size indicates non-random distribution.

The above tests are concerned with detecting departure from randomness, though the variance : mean ratio is also a useful measure of degree of departure. A number of other measures have been proposed which give an indication of degree of departure rather than a test of its significance.

(7) *Index of clumping.* David and Moore (1954) propose the use of

the value $\left(\dfrac{\text{Observed variance}}{\text{Observed mean}} - 1\right)$, which they term *Index of*

Clumping and which is zero for random distributions. They give a test for the significance of the difference between values obtained from two different samples, e.g. for the same species in two different habitats. It depends on the same number of quadrats being used in each case and takes the form of calculating

$$\omega = -\frac{1}{2}\log_e \frac{v_1 \lambda_2}{\lambda_1 v_2}$$

where λ_1, λ_2 are the observed means and v_1, v_2 are the observed variances of the two sets of data. If ω lies outside the range $\pm\ 2\cdot5/\sqrt{(N-1)}$, where N is the number of quadrats in each set, a significant difference in index of clumping is indicated.

If half or more of the total frequency is in the first three classes (0, 1, 2 individuals per quadrat), David and Moore recommend an alternative procedure based on the ϕ test of non-randomness. This alternative procedure, however, tests difference between the means and between the variances of the two sets jointly. Since the means are more likely than not to be different, it is of limited value only.

(8) *Observed density : calculated density ratio.* If a species is contagiously distributed, fewer quadrats will be occupied than in a random distribution of the same density. Thus, if density is calculated from the observed frequency according to the relationship $F = 100(1 - e^{-m})$, the value obtained will be less than the true one. The ratio of observed density to density calculated from frequency will therefore give a measure of the degree of non-randomness, and this will be greater than one for contagious distribution and less than one for regular distribution. A table of expected densities is given in Appendix B (TABLE 6). This measure was apparently first used by McGinnies (1934).

(9) Fracker and Brischle (1944) used a measure essentially similar to the last, calculating the ratio of the difference between the observed and calculated densities to the square of the calculated density, i.e.

$$\frac{m_{\text{obs.}} - m_{\text{calc.}}}{(m_{\text{calc.}})^2}$$

This is zero for random, positive for contagious, and negative for regular distributions. Fracker and Brischle consider, apparently from empirical considerations, that values of from 0·0001 to 0·003 are to be expected for random distributions and that values above 0·02 indicate definite contagion.

(10) *Abundance : frequency ratio.* Whitford (1949) suggested the ratio of abundance (defined as mean density within occupied quadrats) to frequency as a measure of contagion. Abundance is related to density and frequency for

$$\text{Abundance } A = \frac{\text{Total number of individuals}}{\text{Number of occupied quadrats}}$$

$$\text{Density } D = \frac{\text{Total number of individuals}}{\text{Total number of quadrats}}$$

$$\text{Frequency } F = \frac{\text{Number of occupied quadrats}}{\text{Total number of quadrats}} \times 100$$

so that $A \times F = 100D$. The same density may be produced by high frequency and low abundance (regular distribution) or low frequency and high abundance (contagious distribution). The ratio abundance : frequency is equal to $\dfrac{100D}{F^2}$.

This ratio has no fixed expectation for a random distribution, which severely limits its utility in making comparisons. Thus the expected value for $F = 1$ per cent is $1\cdot005$ and for $F = 99$ per cent is $0\cdot047$. Contagious distributions will give a value greater and regular distributions a value less than random expectation.

(11) Morisita (1959a) has developed a measure of departure from randomness based on the measure of diversity proposed by Simpson (1949) rather than directly on Poisson distribution. Simpson's measure is $\sum \pi^2$, where $\pi_1, \ldots \pi_z$ are the proportions of individuals falling into each of Z groups. It ranges from 1 when all individuals are concentrated into one group to $1/Z$ when they are distributed equally between groups. $\sum \pi^2$, the true population measure, is estimated by $\delta = \dfrac{\sum n(n-1)}{N(N-1)}$ where $n_1, \ldots n_q$ are the numbers of individuals observed in q groups and N is the total number of individuals observed. Morisita takes quadrats as the groups and sets up as a measure of dispersion

$$I_\delta = q\delta,$$

which will have the value 1 for random distribution, range from 1 to q for contagious distributions (or more accurately will tend to 1 for random distribution as N increases) and be below 1 for regular distributions.

Morisita proposes to test departure from randomness by referring

$$\frac{I_\delta (N-1) + q - N}{q-1}$$

to tables of F with $n_1 = q-1$, $n_2 = \infty$. The numerator of this fraction is, however, the index of dispersion, $S(x-\bar{x})^2/\bar{x}$, used in (2) above and may be referred directly to the table of χ^2.

For the *Carex flacca* data

$$N = 145, \ N(N-1) = 20880,$$

$$\sum n(n-1) = \sum n^2 - N = 531 - 145 = 386$$

$$\delta = \frac{\sum n(n-1)}{N(N-1)} = 0\cdot0184866$$

$$I_\delta = 3\cdot69732$$

$$I_\delta(N-1) + q - N = 587\cdot41, \text{ as before.}$$

A comparison of the average degree of non-randomness in two or more communities may sometimes be desirable. Measures of non-randomness for individual species might be combined in various ways. Curtis and McIntosh (1950) suggest that the ratio of the sum of observed densities to the sum of expected densities calculated from frequency should be used, i.e.

$$\frac{\Sigma m_{\text{obs.}}}{\Sigma m_{\text{calc.}}}$$

They point out that the measure is weighted in favour of the more plentiful species, which 'seems proper in a general statement about a plant assemblage'.

A number of tests and measures of departure from random expectation are available and it is important to consider their relative value. At first sight it might appear that the χ^2 test of goodness of fit with Poisson expectation would always be the most satisfactory test to use, but it has some disadvantages. Before calculating χ^2 those classes with a low expectation must be pooled; the usual, arbitrary, limit is a minimum expectation of five in each class. One of the commonest effects of non-randomness is the occurrence of a few quadrats with an unusually high number of individuals and, correspondingly, an increase in the number of quadrats with very few individuals or none. It is the quadrats with a high number of individuals and (if the mean number is high) those with few or none which have low expectations and have to be pooled. This may result in the discrepancy from Poisson expectation being largely obscured and the test failing to show any indication of non-randomness. Consider the following data for 100 throws:

Number per quadrat	Observed number of quadrats	Expected number of quadrats	
0	21	14·08	
1	27	27·61	*Mean*
2	22	27·05	*number*
3	14	17·68	*per quadrat*
4	8	8·66	1·96
5	2	3·39	
6	3	1·11	
7	1	0·31	
8	2	0·08	
> 8	0	0·03	

This shows strong indication of contagion and gives a variance : mean ratio of 1·65 (probability from Index of Dispersion of much less than 0·1 per cent). Before applying the χ^2 test the table must be recast:

Number per quadrat	Observed number of quadrats	Expected number of quadrats
0	21	14·08
1	27	27·61
2	22	27·05
3	14	17·68
> 3	16	13·58

which gives a χ^2 of 5·55 with three degrees of freedom and probability 10–20% not indicating departure from randomness. Another difficulty arises when the mean value is low and it is only possible to group the quadrats into two classes, empty and occupied, if all classes are to have an expectation greater than five and there is then no degree of freedom available for a χ^2 test, e.g. 100 throws with a mean of 0·29 have random expectation of 74·8 empty quadrats, 21·7 with one individual and 3·5 with two or more individuals so that the expected numbers must be grouped as 74·8 empty and 25·2 occupied quadrats. Both difficulties may be avoided by increasing the number of samples but it is frequently unpractical to do so without involving labour disproportionate to the information to be obtained.

The variance : mean ratio derives from one particular aspect of departure from Poisson expectation, the occurrence of abnormally high or low variance. Consider the following hypothetical case, for 101 throws with a mean of 1·00, quoted by Evans (1952):

Number per quadrat	Observed number of quadrats	Expected number of quadrats
0	20	37·16
1	76	37·16
2	—	18·58
3	—	6·19
4	—	1·55
5	5	0·31
>5	—	0·05

In spite of the evident non-randomness, the variance is exactly equal to the mean, though the direct test against expectation

69

shows a χ^2 of 63·24, with two degrees of freedom and probability much less than 0·1%. Such a combination of generally regular distribution with occasional groups of individuals is unlikely in natural vegetation and would in any case be recognized without reference to a test. Much the commonest type of non-randomness is the occurrence of excessive number of empty quadrats and of those with high numbers of individuals, owing either to exclusion of the species from part of the area by unfavourable environmental factors or by the presence of competing species, or to aggregation brought about by inefficient propagule dispersal or by vegetative propagation. The variance : mean ratio test is normally sensitive to this type of discrepancy. The last two examples emphasize that either one of the χ^2 test of goodness of fit and the variance : mean ratio test may detect evident non-randomness when the other fails to do so. This has been demonstrated both for artificial 'communities' of discs with known pattern (Greig-Smith, 1952a) and for woody plants in forest (Greig-Smith, 1952b).

The variance : mean ratio as an indicator of non-randomness has been criticized by Skellam (1952) on the grounds that it is dependent on the size of quadrat used. It has been shown earlier in this chapter, however, that the appearance of non-randomness in discrete samples is always dependent on the size of sample unit used. The same supposed disadvantage applies to all the tests and measures proposed, as Curtis and McIntosh (1950) have shown for several of them. It is, indeed, the varying behaviour of non-random distributions with different sample sizes that enables information to be obtained on the scale at which non-randomness is operating. Jones (1955–6) has criticized the variance : mean ratio test on two grounds. He says that it appears to behave erratically when the mean is very small 'presumably because the distributions of deviations of the variance of a Poisson distribution from its mean is too strongly skewed'. It is true that results of the variance : mean ratio test must be interpreted with caution when the mean is very low, but it is under these conditions that a χ^2 test of goodness of fit can be applied only by greatly increasing the sample number. Moore's ϕ test, devised primarily for cases where the mean is relatively high, is also erratic with a very low mean, on account of the low expectation for quadrats containing two individuals. When the mean is very low we are thus dependent on the variance : mean ratio test for lack of a better one. Jones's second criticism is really a more general one, applying to all the tests and measures considered, viz. that it ceases to be applicable with the 'more abundant species' (more accurately those with a high density relative to the maximum possible), where the Poisson distribution itself ceases to be applicable. It has been shown that under these conditions a binomial type of expectation is more appropriate than a Poisson one. A binomial distribution is always more regular than a Poisson distribution of the same mean (cf. TABLE 5). Thus, if the distribution of a species with density high relative to the maximum possible is shown to be contagious when compared with Poisson expectation, it can be assumed that it is contagious when compared with the true random expectation. In the absence of any means of setting up the true random expectation it is impossible to tell whether an observed distribution which is regular compared with Poisson expectation is also more regular than the true random expectation. However, in practice regular distributions are rare in the field and the difficulty is unlikely to occur often.

Moore's ϕ test, like the variance : mean ratio test, is sensitive only to certain types of departure from Poisson expectation but, like it, is usually affected by an undue proportion of empty quadrats, the commonest type of non-randomness. It takes cognizance, however, of less of the distribution than the variance : mean ratio and its principal advantage is in speed of enumeration and calculation.

Of the various measures of non-randomness proposed, the ratio of observed to calculated density, and the indices proposed by Fracker and Brischle and by Whitford all depend solely on the relative number of empty and occupied quadrats. They are thus affected by one aspect only of discrepancy, though an aspect that is frequently the most important in vegetation. Among them Whitford's index has the serious disadvantage that it has no fixed expectation so that only species of the same density *or* the same frequency can be compared directly. There seems no reason for preferring Fracker and Brischle's more complex formula to the straightforward ratio of observed to calculated density. There is no ready means of testing whether any difference between the values of these measures for two sets of data is significant, so that considerable caution must be used in comparing degrees of contagion of different species or in different areas. As a measure of non-randomness the variance : mean ratio is likely to be at least as sensitive as those dependent only on frequency, and in many cases more so. In the modified form of David and Moore's Index of Clumping its variance can be estimated so that the significance of any difference in degree of contagion can be assessed. In general, therefore, it is to be preferred as a measure.

In view of the interest in measures of point-to-plant and plant-to-plant distances as a means of estimating density, it is not surprising that attempts have been made to use a similar approach to detection of non-randomness (Skellam, 1952; Dice, 1952; Hopkins, 1954; Moore, 1954; Clark and Evans, 1954a; Thompson, 1956; Cottam et al., 1957; Pielou, 1959, 1962a). Various suggestions have been made but almost all depend on the equivalence in a random distribution of mean distance from a random point to the nearest individual and mean distance from an individual to its nearest neighbour. One exception is Moore's suggestion that an area should be arbitrarily subdivided and the variances of nearest neighbour distance for the subdivisions compared, in order to test the uniformity of density over the area. Dice (1952) also avoids determination of density. He suggests examining the frequency distribution of the square roots of measurements to the nearest neighbour in each of the sextants around randomly selected individuals. The frequency curve obtained is normal for random distributions but skewed to the right for regular and to the left for contagious distributions. Clark and Evans (1954a) compare the mean nearest neighbour distance with that expected from density. Thompson (1956) has extended this comparison to the mean distance to the nth neighbour. Pielou (1962a) has considered the distribution of shorter nearest neighbour distances only, with the aim of detecting regularity, due to competition, within patches of high density. Pielou (1959) compares the mean distance from random points to nearest individual with that expected from density. (Mountford, 1961, has pointed out that Pielou's procedure fails to take account of the sampling variability of the density estimate.) Skellam's (1952) procedure likewise depends on knowledge of the density. This is a serious objection in practice as the value of the distance measures is greatest in those circumstances where an enumeration of individuals to obtain the density is most difficult. Moore (1954) and Hopkins (1954) have both suggested using, in effect, an estimate of density obtained from point to nearest individual measurements instead. Hopkins uses as a *Coefficient of Aggregation* the ratio of the square of the mean distance between a random point and its nearest neighbouring individual (P) to the square of the mean distance between an individual and its nearest neighbouring individual (I), i.e. provided the same number of measurements are made between pairs of individuals and from points to individuals,

$$A = \frac{\Sigma P^2}{\Sigma I^2}.$$

This coefficient is unity for random distributions, greater than one for contagious distributions, and less than one for regular distributions.

The deviation of A from unity is tested for significance in the following manner. The parameter $x = \dfrac{A}{1 + A}$ has a mean value of 0·5 for random distributions and variance of $\dfrac{1}{4(2n + 1)}$ ($n =$ number of pairs of observations). The distribution of x tends rapidly to normality as n increases and if n is greater than 50, $(x - 0·5)$ may be treated as a normal deviate with zero mean and standard error $\dfrac{1}{2\sqrt{(2n + 1)}}$, i.e. the expression $2(x - 0·5)$ $\sqrt{(2n + 1)}$ may be referred to standard tables of the normal integral. If n is less than 50 reference must be made to tables of the incomplete beta function (Pearson, 1934). Hopkins gives a diagram

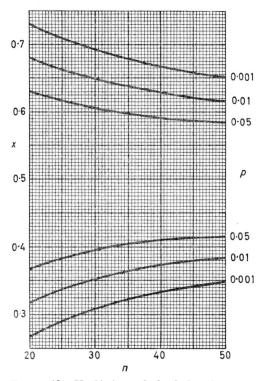

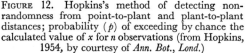

FIGURE 12. Hopkins's method of detecting non-randomness from point-to-plant and plant-to-plant distances; probability (p) of exceeding by chance the calculated value of x for n observations (from Hopkins, 1954, by courtesy of *Ann. Bot., Lond.*)

(FIGURE 12) showing the value of x corresponding to probabilities of 5,1 and 0·1 per cent and values of n from 20–50. There is one point of practical importance. Often the only practicable method of selecting a 'random' pair of individuals is to take the nearest individual to a random point and measure the distance to its nearest neighbour. As Cottam and Curtis (1956) have shown, this results in a non-random sample giving a mean distance greater than the true value. Thus the expected values of A may be less than 1 and caution must be exercised in accepting a distribution as regular.

Cottam *et al.* (1957) have attempted to interpret the meaning of indices of non-randomness derived from distance measures. They point out that, taking the simplest case of grouping of individuals into distinct clumps, aggregation can be described in terms of three variables, the clump mean area (total area of the population divided by the number of clumps), area of clumps (the mean value of the areas that would be obtained by drawing a planimeter round each clump) and the within-clump mean area (the mean value for the area of a clump divided by the number of individuals it contains). If any two of these factors are known for a fixed number of individuals in a fixed area, the third can be determined. Plant-to-plant distance is determined by within-clump mean area. It is this dependence on within-clump mean area that limits the value of indices based on plant-to-plant distances. Such indices will reflect mainly the nature of the clumps, primarily a result of reproductive behaviour and generally obvious on inspection, and scarcely at all the biologically more interesting pattern of the clumps themselves. An index based on point-to-plant distance is affected by the pattern of clumps as well, as Pielou (1959) has recognized, and is to that extent more satisfactory.

Clark and Evans (1954b) have shown that there is a fixed expectation for a random distribution of the proportion of individuals serving as nearest neighbour to 0, 1, 2, 3, 4, and 5 others. This proportion is affected by departure from randomness only if it involves direct effects between individuals. Clark (1956) has derived for a random distribution the proportion of individuals for which the relationship with the nth nearest neighbour is reflexive, i.e. each of the pair is the nth nearest neighbour of the other. Observed distributions may be compared with this random expectation. The labour involved in collecting data makes it unlikely that these approaches will have any great practical value. Bray (1962) has suggested another approach independent of distances measured and of any estimate of density. If n individuals are recorded at each sampling point, then, if a species is randomly distributed, the number of points at which 0, 1, 2, ... n individuals of that species are recorded will fall on a binomial distribution with p equal to the observed proportion of the species in all individuals recorded. A χ^2 test may be used to test the observed distribution against the expected binomial distribution.

The methods of detecting and measuring departure from randomness so far discussed are applicable only to species occurring as discrete, readily distinguished individuals, or those which present other countable units, e.g. rosettes or tillers. The latter usually show pronounced contagion owing to a number arising close together from the same individual and it is difficult to distinguish this from any non-randomness of individuals present. In many species even this modification of counting rosettes or tillers is impossible, e.g. many procumbent, rooting herbs. In many communities, therefore, it may be impossible by these techniques to analyse the distribution of all, or even of a majority, of the species present.

Species not amenable to counting, and many species that are, can conveniently be measured in terms of cover. If a set of points linked together in a constant manner, e.g. the frame of ten pins commonly used in making estimations of cover, is placed on vegetation in which the distribution of cover is random, the chance of a set of n points showing presence of a species at 0, 1, 2, 3, . . . n points will be given by expansion of the binomial series $(p + q)^n$ where p is the mean cover and $q = 1 - p$. If cover is contagiously distributed, i.e. if there are patches within which cover is greater than the mean for the whole area, then there will be a greater number of sets with no, or few, points occupied and with a high number occupied. Conversely, if cover is more uniformly distributed than random there will be a greater number of sets with about the mean number of points occupied.

The concepts of contagious and regular distribution in relation to cover should perhaps be examined more closely. In the density approach to pattern the convention is maintained that individuals occur at a point. It is because this is not true that the Poisson expectation for random distribution holds only within certain limits and we are forced to introduce the idea of density relative to the maximum possible. In considering cover no such simplifying convention is possible. Consider first a pair of points set a known distance apart as the sampling unit. The smallest unit of cover is a leaf or a length of stem. If the distance between the points is less than the average dimension of a leaf, the points will most commonly either both touch a leaf or neither touch a leaf and the data will show contagion when compared with random expectation. If the leaves are themselves uniformly distributed (cf. for example the circles in FIGURE 13), and the two points are about the same distance apart as the mean distance between the leaves, an undue number of pairs will tend to fall one on a leaf and one off, i.e. a regular distribution will be indicated. If the points are at very much greater distance apart than this, then the observed data will tend to random expectation. In practice we are not likely to work with point spacings anywhere nearly as small as the dimensions of leaves, but the same considerations apply to spreading individuals or patches of species as apply to leaves, except that the indications will not be as clear-cut because there is not normally complete cover within the boundaries of one individual. To offset this the number of points within a set may be increased, the distance between the

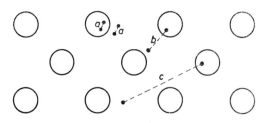

FIGURE 13. Effect of spacing of sample points on indications of non-randomness of cover. (a) Spacing indicating contagious distribution. (b) Spacing indicating regular distribution. (c) Spacing indicating random distribution (see text)

extreme members of the set bearing the same relation to the pattern of cover as the distance between the two points in the hypothetical case we have been discussing. It is evident from these considerations that, just as with density data, indications of non-randomness obtained will depend on the size of sampling unit used.

The proportions of sets with various numbers of points occupied may be compared with the binomial expectation for the same mean by a χ^2 test of goodness of fit, just as density data are compared with Poisson expectation, e.g. 100 throws of a frame of four points arranged at the corners of a square of side 15 cm, were made on *Alchemilla vulgaris* agg. in a lawn. The mean cover was 14·75 per cent. The observed number of frames having 0, 1, 2, 3, and 4 points, and the number expected from the expansion of $(0·8525 + 0·1475)^4$ were

Number of points occupied	Observed number of frames	Expected number of frames
0	64	52·82
1	20	36·56
2	10	9·49
3	5	1·09
4	1	0·05

After pooling values for 2, 3 and 4 points occupied, χ^2 is 12·58, with one degree of freedom and probability less than 0·1 per cent indicating departure from randomness.

A test comparable to the variance : mean ratio used for density data may also be used. The variance of a binomial series $(p + q)^n$ is npq and the sampling variance of this variance is

$$\frac{2(npq)^2}{N - 1} + \frac{npq\,(1 - 6\,pq)}{N}$$

where N is the number of throws (Fisher, 1941, § 18). Variance of the observed data may thus be compared with that expected. For the example quoted

$$Sx = 59,\ Sx^2 = 121,\ S(x - \bar{x})^2 = 121 - \frac{59^2}{100} = 86\cdot19 \text{ and}$$
$$\text{Variance} = \frac{86\cdot19}{99} = 0\cdot8706.$$

The expected variance is $npq = 4 \times 0\cdot1475 \times 0\cdot8525 = 0\cdot5029$. Difference between observed and expected variance is 0·3047. Sampling variance of variance

$$\frac{2 \times 0\cdot5029^2}{99} + \frac{0\cdot5029\,(1 - 0\cdot7544)}{100} = 0\cdot006345.$$

Standard error of variance $\sqrt{0\cdot006345} = 0\cdot07966$.

The ratio of the difference to its standard error $\dfrac{0\cdot3047}{0\cdot07966} = 3\cdot83$

with probability, from tables of the normal integral, of about 0·01 per cent. Alternatively, as with the Poisson distribution, the index of dispersion may be referred to the table of χ^2. The form of the index of dispersion appropriate to the binomial distribution is $\dfrac{S\,(x - \bar{x})^2}{\bar{x}q} = \dfrac{S\,(x - \bar{x})^2}{npq}$. For the example quoted this is $\dfrac{86\cdot19}{0\cdot5029}$ $= 171\cdot39$, with 99 degrees of freedom and probability of less than 0·01 per cent.

A useful approximate measure of departure from randomness, comparable to the ratio of observed to calculated density used for density data, is the ratio of the observed cover to cover calculated from frequency (percentage of frames with at least one point occupied). The expected proportion of frames of n points with no points occupied is $(1 - p)^n$. Hence the frequency $F = 100 - 100 (1 - p)^n$ and from this the cover corresponding to any given frequency can readily be calculated. In the present case the observed frequency is 36 per cent and $100 (1 - p)^4 = 64$ from which $p = 0.1056$ and the ratio of observed to calculated cover is

$$\frac{0.1475}{0.1056} = 1.40.$$

If the mean cover is very low only a limited number of possible distributions of occupied points can occur, e.g. for 100 throws of four points with a mean cover of 1 per cent, only five arrangements are possible. Under these conditions the use of tests based on continuous distributions is not valid. (The same considerations would apply if percentage cover approached 100 per cent but this is unlikely to occur in practice.) The difficulty can clearly be overcome by increasing the sample size, but this may not be practical. An alternative is to replace the sampling points by quadrats of suitable size, recording presence or absence only of the species concerned. The simplest method in most cases is to use a group of contiguous quadrats as the sampling unit. The mean frequency obtained in this way will be higher than the mean cover and, since the data may be treated in exactly the same manner as that from points, a suitable size of quadrat will permit tests of non-randomness on species too sparse to be tested by cover.

Another approach depending on frequency has been suggested by Jones (1955–56). If a number of quadrats are arranged in either a 'chessboard' pattern or a line, the occurrence of contiguous pairs or runs of quadrats containing a species may be examined. If aggregation of a species is on a scale larger than the size of quadrat used, the species will tend to occur in adjacent quadrats more frequently than would happen if distribution were random. If departure from randomness is on such a small scale that a unit of the resulting pattern is contained within one quadrat, then it will not be detected. On the other hand a widespread and abundant species may have such a high frequency in quadrats of the size used that departure from randomness may not be detected, although clearly demonstrable by density analysis. Formulae for calculating the expected number of contiguous pairs and runs under various conditions and the variances to which they are subject have been given by Krishna Iyer (1948, 1950).* The difference between the observed number of contiguous pairs or runs and that expected, compared with its standard deviation, is referred to tables of the normal curve.

In the opening paragraphs of this chapter it was argued that the biological significance of any departure from randomness lies in the indication it provides that one or few factors are playing an overriding part in determining the survival of individuals. Departure from randomness is thus not of great interest in itself; its importance is as a guide to the factors controlling the survival and performance of individuals. Various attempts at a generalized interpretation of non-randomness have been made. It has frequently been pointed out that the commonest source of discrepancy between observed data and random expectation is the excessive number of blank quadrats, with the variance exceeding the mean. David and Moore (1954) state the issue succinctly: ' "Student" (1919) pointed out that the variance being greater than the mean is the usual cause of the Poisson hypothesis being inadequate for field data, and two mechanisms have since been put forward to "explain" this inadequacy. The first mechanism assumes that the Poisson parameter, λ, varies over the field. In the second mechanism the Poisson parameter is assumed to remain constant, but there is some form of dependence among the observations, a concept which Pólya (1930) called "contagion". . . . Feller (1943) has shown how it is impossible to distinguish between these two mechanisms on the basis of an observed set of data,† and probably the true state of affairs lies in between the two and may well be different for each species'.

It is evident that both mechanisms modifying the Poisson hypothesis are operative in natural vegetation, and the last phrase quoted is certainly an understatement. Vegetative spread and, for many species, seed dispersal will lead to the second mechanism being effective. Any botanist with field experience is well aware of the widespread occurrence of point-to-point variation within a plant community of such scale and degree that it is apparent by qualitative criteria alone. Watt (1947) has shown that some cases of such

* The most generally useful formulae are those for a line of quadrats under conditions of 'non-free sampling', i.e. using the observed frequencies to calculate the expected values. The expected number of joins between quadrats with and without the species is $\dfrac{2\,n_1\,n_2}{m}$ and its variance is $\dfrac{2\,n_1\,n_2\,(2\,n_1\,n_2 - m)}{m^2\,(m-1)}$, where n_1, n_2 are the number of quadrats respectively containing and not containing the species, and $m = n_1 + n_2$ is the total number of quadrats in the line.

† More precisely, Feller showed it to be impossible *on the basis of an overall observed frequency distribution*.

variation can be explained by the occurrence of cyclic changes in the composition of the vegetation, changes determined directly by changing behaviour of individuals with age, or indirectly by reaction of different species on their immediate environment. Other cases are dependent on predetermined differences in soil, topography or other environmental factors, varying in scale from the correlation of vegetation with catenas of soil types to the difference in composition of vegetation between ridge and furrow in grassland on old arable land. These are gross differences correlated with obvious environmental features, but it is evident that similar differences, of lesser degree, will occur, being expressed rather as differences in amounts of species and detectable only by quantitative means.

In spite of the abundant evidence that both mechanisms are likely to be operative in producing deviation from Poisson expectation, as was pointed out by Clapham (1932), attention has been very largely focused on the presumed consequences of reproductive behaviour. Attempts to account for all non-randomness on this basis alone are doomed to failure. Nevertheless it is worth while to examine various suggested distributions and consider their adequacy when point-to-point variation is absent.

1. Archibald (1948) tested data for various salt-marsh species against Neyman's Contagious distribution and found a reasonably good agreement for species that did not fit the Poisson distribution. Fracker and Brischle (1944) on the other hand found that Neyman's distribution did not fit data for two species of *Ribes* in forest and suggested that there was a mixture of random original immigrants and contagiously distributed later generations. Neyman (1939), concerned with recently hatched insect larvae crawling away from egg clusters, assumed random distribution of larval clusters with the number per cluster also random, but with a limit to the distance to which larvae might crawl, i.e. an arbitrary limit to cluster size. His distribution is defined by two parameters, one proportional to the mean number of clusters per unit area and one to the mean number of units per cluster.

2. Archibald (1950) tested a number of species from several communities (evidently principally from salt marsh and chalk grassland) against Thomas's Double Poisson distribution (Thomas, 1949) and again found that many species fitted it well. This distribution is based on similar assumptions to that of Neyman, clusters being assumed to be randomly distributed and the number of units, additional to the first, also being random.

Barnes and Stanbury (1951) found that data for early colonization of newly-exposed surfaces of china clay residues could be fitted to both Neyman's and Thomas's distributions. (Pielou, 1957, has given reasons for expecting the two distributions to be fitted equally well by quadrat data.) Thomson (1952), on the other hand, found that only one species out of three tested in an 'old-field' community could be fitted to either distribution.

Evans (1953) has discussed the data of Archibald and Barnes and Stanbury in more detail and concludes that Neyman's distribution gives a satisfactory fit.

3. Robinson (1954) has suggested the use of the Negative Binomial distribution. If groups are randomly distributed and the number per group follows a logarithmic distribution then the resultant number of individuals per quadrat is given by the negative binomial distribution. Robinson fitted data of Steiger (1930) on prairie vegetation to this distribution and found good agreement. Evans (1953), however, considered the fit to the negative binomial not satisfactory. Unfortunately Steiger's data are scarcely adequate material for a test of the distribution, for, as Curtis (1955) has pointed out, Steiger deliberately selected the position of his quadrats to include as great a range of densities as possible and they cannot be considered unbiased samples for this purpose.

Various other distributions of a similar type have been suggested, but none of them, apparently, has been tested against field data.

Neyman's and Thomas's distributions are based on such similar assumptions that they may be considered together. Even if the primary assumption that reproductive behaviour is mainly responsible for non-randomness is accepted, the assumptions made in applying the distributions are not free from objections, which have been well put by Goodall (1952a): 'Both are open to the objection that the range of each postulated group is limited—in one case arbitrarily, in the other to the observational quadrat. A more natural assumption would be that the probability of finding a plant falls off as some continuous function of the distance from the centre of the group of which it is a member—the number of propagules being perhaps related to distance from mother plant by an inverse square law.' A further objection is the absence of any real evidence that the number of individuals per cluster is distributed randomly. Where spread of a species is vegetative this is probably often not so, as study of morphology indicates in many cases a tendency to production of a constant number of daughter individuals per individual per season. The successful fitting of either distribution provides, from the second parameter, an estimate of the number of individuals per clump. This is of some intrinsic interest but it is little help in disentangling the biological relations of the species of the community with one another and with environmental factors.

The use of the negative binomial is likewise dependent on an unproved assumption on the nature of clusters. The distribution is defined by the mean number per quadrat and by the exponent, which is dependent on the mean reproductive rate. Here again the utility of an estimate of the second parameter is doubtful. Robinson claims on theoretical grounds that the value of the exponent should be independent of external conditions and hence, for the same species, constant in different communities. He draws support from Steiger's data for this. It is not clear why reproductive behaviour should remain the same under different conditions; indeed, general knowledge of plants in the field would suggest that it would not do so. Little weight can be attached to Steiger's data in this connection for reasons already considered. Further, the thesis of a unique value of the exponent for any one species is not supported by Pidgeon and Ashby's (1940) demonstration that species may be randomly distributed in one area and markedly contagious in another.

In comparing the various suggested distributions, it must be remembered that there are so many possible theoretical distributions of this type that the satisfactory fit of data to one of them can scarcely be regarded as evidence for its validity in the absence of strong and independent biological evidence for the truth of its underlying assumptions.

The importance of point-to-point variation is emphasized by the fact that satisfactory fit to Neyman's and Thomas's distribution has only been obtained under conditions where such variation is minimized. Archibald's data are derived from samples of 100 or 500 *contiguous* quadrats each of area 20 sq. cm so that even with samples of 500 the total area sampled was only one square metre, small enough to avoid most if not all of any point-to-point variability present, particularly in the relatively uniform habitats of salt marsh and chalk grassland sampled. Barnes and Stanbury deliberately worked on an exceptionally uniform habitat (residues from china clay mining allowed to settle behind artificial dams) and were concerned with early stages of succession carried on by water- and wind-borne migrules.

The only mathematical approach to non-randomness based on the first mechanism, that of variability of the Poisson parameter over the field or point-to-point variability, appears to be that of Stevens (1937 and in Ashby, 1935). He derived the formula $E = n\left(1 - \dfrac{1}{n}\right)^s$ for the number of empty squares expected in a quadrat divided into n smaller squares and containing s individuals randomly distributed (p. 65). If the distribution is not random he suggested substituting

$$E = n\left(1 - \frac{1}{n}\right)^{s}[1 + s\,(s - 1)\,c]$$

where c is calculated to give the best fit to the data. c is then a measure of the variability of the Poisson parameter over the field. This is applied to *Salicornia europaea* agg. in salt marsh by Ashby (1935) and to *Bonnaya brachiata*, a weed species on arable land in India, by Singh and Das (1938). In neither case was any test made of goodness of fit given by the corrected formula for E, but in each case it clearly accounts for a considerable part of the departure from randomness. Although this approach is based on point-to-point variability, in both cases where it has been tried the sample quadrat used was of such size that only small-scale heterogeneity, due to clusters of the order of size produced by reproductive behaviour, would be likely to be detected. In other words, although the first mechanism (of point-to-point variability) was assumed, the second (of reproductive behaviour) was in fact investigated. Feller's (1943) conclusion that it is impossible to distinguish the effects of the two mechanisms is once more emphasized. Theoretically there is no reason why the sampling quadrat should not be made very much larger to take account of point-to-point variability, but, except for species of very low density, this would involve so many squares within the quadrat to allow a reasonable proportion of empty squares that enumeration of a sufficient number of quadrats would be impractical. There is also a limitation in that the method applies only when c is small relative to the mean probability of occurrence of an individual. Finally it is difficult to interpret the value of c either in terms of the plant or of the habitat.

Erickson and Stehn (1945) suggested that a given habitat could be divided into two parts, one favourable to a species and the other unfavourable or at least not occupied, and that observed data could be divided into two sets, one derived from the favourable and one from the unfavourable portions of the habitat. To obtain such a division they plotted log $(x!y)$ against x (y being the number of quadrats containing x individuals). A Poisson distribution plotted in this way gives a straight line

$$\log (x!y) = \log (Ne^{-m}) + x \log m$$

(from $y = (Ne^{-m}m^{x})/x!$, for N quadrats with mean m). They fitted a straight line by eye to the points corresponding to higher values of x and from it derived the number of, and mean density within, quadrats to be assigned to the favourable portion of the area. The distribution for quadrats assigned to the unfavourable portion was then obtained by difference and, in the majority of cases tested, was found to give a good fit to Poisson distribution. This approach, which has been largely overlooked, is realistic in concentrating attention on variability in the Poisson parameter. Its value is, however, limited by the assumption that the Poisson parameter shows discontinuous variability; this is unlikely in many circumstances. It will be invalidated if clumping due to reproductive behaviour is present.

Kemp and Kemp (1956) have considered the expected distribution when cover is recorded by frames of pins. Following Robinson's (1954) suggestion that cover might be expected to follow the Beta distribution,* they derive a distribution based on the assumption that the probability of a pin hitting the species is constant within a frame but has a Beta distribution between frames. The resulting distribution may be defined in terms of the mean and variance of the Beta distribution. The variance of the Beta distribution is a measure of the variability between frames, i.e. of point-to-point variability. They obtained a satisfactory fit to cover data taken from Goodall (1952b). Once again the basic assumptions may be questioned. No real evidence has been presented for the appropriateness of the Beta distribution. Further it is unlikely, for the usual spacing of pins, that the probability within a frame is constant, though this objection may be overcome by using a smaller frame. Lastly, as with Stevens's approach, the second parameter, though measuring point-to-point variability, is difficult to interpret in biological terms.

Summing up the results of attempts to use modifications of the Poisson or binomial distributions to account for discrepancies found in field data, it appears that they result either in a statement of the obvious, viz. that a species is tending to occur in compact clumps of several individuals, or in definition of the pattern of distribution in terms of constants to which no precise biological meaning can be attached. It seems likely that a more empirical approach, while inevitably unattractive to the mathematical mind, will be more rewarding to the ecologist, who is aware of the very large number of variables which he has to consider in all but the most favourable circumstances.

Two conclusions that have emerged from the various investiga-

---

* Defined by two parameters l and m. The probability of a fractional area x occurring in a quadrat is given by $y = \dfrac{\Gamma\,(l + m + 2)}{\Gamma\,(l + 1)\,\Gamma\,(m + 1)}\,x^l(1 - x)^m$.

tions of pattern so far made are that species are relatively rarely randomly distributed even within small and apparently homogeneous areas, and that contagious is very much more common than regular distribution (Ashby, 1948). The rarity of regular distribution is at first sight surprising, as it might be expected to occur as the result of competition between individuals. Absence of reports of regular distributions might perhaps be attributed to many workers having selected the easier species in a community, i.e. normally the less dense, to investigate. However, in forest communities in Trinidad Greig-Smith (1952b), considering all woody species, found only very slight indications of regular distribution even when individuals of different species were grouped under size classes. Similarly, Jones (1955–6) working in forest in Nigeria makes no mention of the occurrence of regular distributions. Though evidence of the relative rarity of regular distributions cannot yet be regarded as conclusive, we are clearly justified in concentrating attention on contagious distributions at the present stage.

If distribution is contagious then, whatever the exact nature of the contagion, the actual occurrence of individuals on the ground is, in qualitative terms, patchy; either there are patches in which a species is present and patches in which it is absent or patches in which it occurs more abundantly and patches of less abundance. If we are to discover the causes of this patchiness information about the size of patches is desirable and may be essential. It will at least permit the elimination of factors which do not exhibit patchiness of a similar scale as possible causes. The distribution of individuals may be more complex than the occurrence of patches of two types, e.g. a species might show groups of plants each originally derived from a single individual by vegetative reproduction, with the groups themselves being more numerous in some patches than in others. In other words heterogeneity may be present on several different scales concurrently and it is desirable to detect the existence of the different scales and their approximate dimensions.

Some indication of scale of heterogeneity may be obtained by varying the size of sampling unit in the tests already described and noting the size at which indications of non-randomness disappear or decrease markedly, but this is a laborious and insensitive method. It is in any case of little practical use for the larger scales generally produced by variation in environmental factors. If the non-randomness is very marked, systematic sampling (e.g. by a grid of contiguous quadrats) and subsequent insertion on a plan of *isonomes* (i.e. lines joining samples of equal density or other measure) (Pidgeon and Ashby, 1942), may be sufficient to indicate relatively large-scale heterogeneity. Greig-Smith (1952a) has suggested a more discriminating technique based on systematic sampling. A grid of contiguous quadrats is used, each side of the grid having a number of grid units which is a power of 2, e.g. 16 × 16, 32 × 64, and the number of individuals per quadrat or grid unit is counted in the normal way. The total variance between grid units can be apportioned by the usual technique of analysis of variance to differences between blocks of 2, 4, 8, 16, etc., grid units, the blocks being square for even powers of two and oblong for uneven, and a remainder representing differences between individual units within blocks of two units, or what may conveniently be termed between 'blocks of one unit'. If the arrangement is perfectly random over the whole area, the mean square for all block sizes should be the same and equal to the mean number of individuals per grid unit, i.e. the variance : mean ratio should be unity whatever sample size is used. If the distribution is contagious it is to be expected (M. S. Bartlett quoted in Greig-Smith, 1952a), and has been found by trial on known patterns of discs, that the variance will rise up to a block size equivalent to the areas of patches. If the patches are themselves random or contagious the variance will be maintained at this level with increasing block size. If the patches are regular, and this is necessarily the case when alternating low and high density areas of similar size are involved, the variance will fall as block size is increased further. If more than one scale of heterogeneity is present the behaviour will be repeated as block size reaches this scale. After analysis of variance of the data obtained from the grid in this way the mean square is plotted against block size and the position of sharp rises and peaks noted.

The testing of significance presents considerable difficulties. As the technique was originally proposed the mean square at higher block sizes was compared with that at unit block size by the variance ratio or F test. A modification of this is to compare adjacent points on the curve in the same way. However, as Bartlett (personal communication) has pointed out, once the existence of non-randomness has been proved, the use of the F test is no longer valid. Thompson (1955, 1958) has examined the problem and shows that if some likely model (in the mathematical sense) of the community being investigated is set up, the expected mean square and its standard error can be calculated for each block size. If the observed mean squares do not deviate significantly, the model can be regarded as a satisfactory description of the pattern. The necessity for a model makes this admittedly much more satisfactory test impractical in most circumstances. We are, therefore, forced back on a subjective judgment of peaks, though the F test is of negative value in indicating differences which are certainly not significant. In practice, assessment of peaks is very materially assisted by their consistency in a series of analyses; if a peak recurs in the same position or shows a regular drift in a series of related communities there can be little doubt of its validity. Kershaw (1957a), in an extensive series of analyses of artificial communities, has demonstrated the general reliability of this subjective assessment. The residual mean square, for blocks of one unit, may, if necessary, be tested against the mean in the same way as data from random quadrats (p. 58). The departure of the ratio from unity has a standard error $\sqrt{\dfrac{2}{N}}$ where N is the number of degrees of freedom appropriate to blocks of one unit.

Thompson (1958) has supported the use of consistency between graphs as a basis for judgment of significance. He has also pointed out that significance bands at any desired level can readily be calculated for a random distribution. If the distribution has unit variance, a sum of squares based on N degrees of freedom is distributed as χ^2_N. Thus for, say, 95 per cent probability bands, the upper limit U will be obtained from the value of χ^2 corresponding to $P = 0.025$ and the lower limit L from the value corresponding to $P = 0.0975$. The limits for the mean square are then U/N and L/N. Such significance bands are particularly useful when only a small number of analyses are available. Thompson provided a brief table of 95 per cent limits and a fuller table from Greig-Smith (1961c), including all values of N that are likely to be used, is reproduced in Appendix B (TABLE 7). The limits are expressed in terms of unit variance, i.e. for density data, as variance : mean ratio.

A considerable improvement in the technique may be made in a way indicated by Kershaw's (1957a) application of the same approach to frequency and cover. Instead of a grid of quadrats a number of parallel lines of quadrats is used and the counts from each line grouped into blocks of successively larger size. Peaks on the mean square : block size curve then indicate the linear dimensions of the patches instead of their area. This gives a considerable saving in the number of quadrats enumerated to detect heterogeneity at any particular scale. Furthermore, if there is any reason to suspect that patches are not isodiametric, as on sloping ground, separate analyses may be made in two directions at right angles. Kershaw points out that the variance of the largest block size (the full length of the transects) must be ignored as it is liable to include an element dependent on the spacing between them.

Frequency may be used instead of density as the measure if each grid unit is subdivided into a number of smaller units in each of which presence or absence is noted, e.g. a grid unit may consist of 25 smaller squares so that any integral value from 0 to 25 may be recorded. The data are then treated in exactly the same way as density data. Use of frequency has the advantage that any species may be enumerated but it suffers from a serious drawback in the relatively large size of grid unit necessary if it is to include a number of sub-units separately recorded. Kershaw (1957a, b) has extended the method to measurements of cover by using line transects, presence or absence being recorded at points equally spaced along the line. Using points 1 cm apart, this has proved satisfactory for grassland, which is intractable to pattern analysis by most methods. The basic unit is a set of five consecutive points on the line. The mean square between blocks of one unit in both frequency and cover analyses may be tested against the variance expected from binomial expectation by an F test (with $n_1 =$ number of degrees of freedom appropriate to the mean square, $n_2 = \infty$). It was at first considered that such data should be subjected to angular transformation before analysis but reconsideration has indicated that, provided tests of significance are not required, there is little to be gained by doing so (Greig-Smith, 1961c). If significance bands are required, however, then transformation of the data is necessary. The expected variance on the transformed scale is approximately $\dfrac{820 \cdot 7}{n}$ (where $n =$ number of points or squares in the basic unit) but this deviates widely from the true value when n is small (Chapter 2, p. 41). The true value can only be obtained by calculation from the corresponding binomial expansion, with the transformed values substituted for the number of points occupied. The variances on the transformed scale for the most usual values of n and a selection of values of mean cover are given in Appendix B (TABLE 1).

Kershaw (1957a) has used artificial 'communities' to examine the sensitivity of this cover method at low values of percentage cover. He concludes that the use of a basic unit not more than half the size of the smallest scale of heterogeneity likely to be present greatly increases sensitivity. If sample size, i.e. total number of grid units, is inadequate, the peaks tend to drift one block size to the right, owing to patches overlapping two blocks of their own size. This drift can be cancelled by an increase in sample size. As cover falls, sensitivity, not unexpectedly, falls also, until at very low cover values the necessary sample will be so large as to make the method impractical. It is impossible from the rather formalized conditions of artificial communities to give a precise estimate of the cover value below which the method should not be used, but it appears that it may probably be used for cover down to 10 per cent and perhaps to 5 per cent, provided that it is possible to use a basic unit sufficiently small to give maximum sensitivity. If cover value is lower than this, frequency must be used, in spite of its disadvantages.

Analysis of pattern by this method is most useful in vegetation that, apart from relatively small scale clumping due to reproductive features, is apparently homogeneous. Sometimes, in spite of appar-

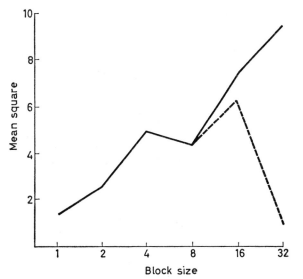

FIGURE 14. Mean square: block size graph for local frequency in 10 × 5 cm units of *Agrostis stolonifera* in a dune slack with (broken line) and without (continuous line) correction for covariance with position (from Greig-Smith, 1961c, by courtesy of *J. Ecol.*)

ent homogeneity, there is in fact a trend in abundance of species along the length of the transects. This results in a steady rise at the larger block sizes which may mask scales of pattern present (FIGURE 14). Any such trend will be apparent if the totals for the larger block sizes are examined, especially if the transects are made $2^{x-1} \times 3$ in length instead of 2^x so that three values for the largest block sizes are available from each transect. The effect of the trend on the mean square : block size graph may be reduced sufficiently to expose smaller scales of pattern by deducting terms for covariance with position from the sums of squares (Greig-Smith, 1961a,c).

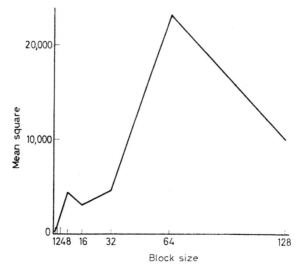

FIGURE 15. Mean square: block size graph for *Agrostis tenuis* in reseeded upland pasture, seven years after sowing. Transformed cover data from transects with basic unit of 5 cm (data from Kershaw, 1957b)

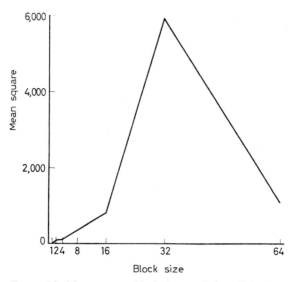

FIGURE 16. Mean square: block size graph for soil depth at intervals of 10 cm along transects in reseeded upland pasture (data from Kershaw, 1957b)

90

The interpretation of these grid analyses may be made clearer by two examples. Kershaw (1957b, 1958) using cover in basic units of 5 cm found consistent peaks at block size 64 (*ca.* 3m) for several species in a series of upland reseeded pastures of different ages. FIGURE 15 shows a typical curve, for *Agrostis tenuis*.* A similar transect analysis of depth of soil at successive points, using a basic unit of 10 cm, gave a peak at block size 32, indicating the same scale of heterogeneity (FIGURE 16). This suggested a correlation of depth of soil and abundance of the species not previously suspected. On following up this indication the relationship was confirmed.

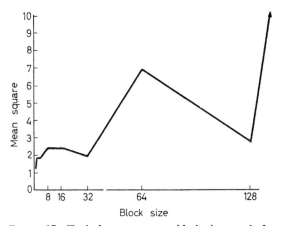

FIGURE 17. Typical mean square: block size graph for *Eriophorum angustifolium* (from Phillips, 1954a, by courtesy of *J. Ecol.*)

Phillips (1953, 1954a) has used the method to obtain information on morphological response of *Eriophorum angustifolium* otherwise available only with great difficulty. Rhizomes arise from the base of upright shoots of *E. angustifolium* in three vertical planes only, owing to the one third phyllotaxy of the species. These may be arranged, relative to the direction of the rhizome of which the parent shoot is the terminal portion, as two forward and one back or one forward and two back, i.e. there are six possible positions which daughter rhizomes may occupy. Those in the three forward positions are longer than backward ones on the same plant. The daughter rhizomes turn up to form new aerial shoots and the same pattern is

* The peak at block size 8 corresponds to smaller patches, due to vegetative spread, within the larger ones. Such 'morphological peaks' commonly appear in analyses, but their nature is generally readily recognized.

repeated. The aerial shoots flower only after several, commonly three, years, after which they, and the rhizomes from which they arise, die. The plant thus consists of a system of limited extent of shoots linked by rhizomes. These facts could readily be established by excavation in suitable open habitats such as mud pools, but the difficulty of extracting the rhizome systems intact from more compact soils prevented direct study of varying length and production of rhizomes in different habitats. A typical grid analysis (FIGURE 17) from a favourable habitat shows a 'primary double peak' (block sizes 2 and 8) corresponding to the grouping around the parent shoot of shoots from the shorter backward and longer forward rhizomes respectively, and a 'secondary peak' (block size 64) due to the grouping of shoots linked in one system by rhizomes. FIGURE 18

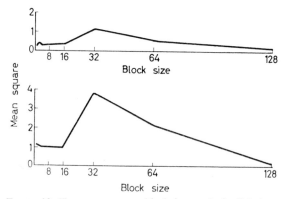

FIGURE 18. Two mean square: block size graphs for *Eriophorum angustifolium* in *Calluna vulgaris-Erica cinerea* communities (from Phillips, 1954a, by courtesy of *J. Ecol.*)

shows two grid analyses from a habitat unfavourable to *E. angustifolium* and illustrates the type of information obtained. Here both peaks occur farther to the left, indicating that rhizomes are shorter, and the primary peak is much lower and not distinguishable into two parts, showing that fewer rhizomes are produced from each shoot.

The relative height of peaks clearly reflects the intensity of the pattern, i.e. it corresponds to measures of non-randomness found from random samples. Intensity of pattern is of some interest as it, in turn, reflects the degree of control by the influencing factors. High peaks indicate rigid control, with a very much greater representation of the species where the controlling factor is favourable than where it is unfavourable. Conversely, low peaks indicate relatively small differences. The width of the peak is determined by the variability of size of patches and is generally of less interest.

The basis of this method of analysing pattern has recently been questioned by Goodall (1961), who claims that the graphs can be explained in terms of an increase in variance with block size over the whole range of block sizes and that the peaks are due to random fluctuations. Greig-Smith *et al.* (1963) have, however, shown that the logarithmic transformations which Goodall has applied to both mean square and block size before fitting a regression are inappropriate to a test of significance of peaks, that the data presented by Goodall show evidence of overall trends and that his test of proportion of graphs having various numbers of peaks is not the appropriate one.

Morisita (1959a) has suggested an essentially similar approach to the analysis of pattern. His measure of dispersion I_δ is calculated for a series of quadrat sizes and the ratio $I_{\delta(s)}/I_{\delta(2s)}$ plotted against quadrat size (taken as $2s$). The resulting graph shows peaks at quadrat sizes corresponding to the scales of pattern present. Morisita has applied this method to the data of Evans (1952) and a more extensive trial in various types of forest has been made by Ogawa *et al.* (1961).

These methods are so sensitive that they can profitably be used only within areas that are apparently homogeneous or nearly so. More marked pattern is often present even within the limits of a community as accepted by most ecologists. Anderson (1960, 1963) has shown that such pattern may usefully be examined by means of the ordination techniques which have been developed primarily in relation to community differences. These techniques are discussed in Chapter 7.

CHAPTER 4

ASSOCIATION BETWEEN SPECIES

IF one or few factors have a predominating influence on the occurrence of individuals of a species, the effect on the spatial distribution of individuals within the community will, as we have seen, depend on the distribution of different levels of the influencing factor. Suppose, for example, the occurrence or non-occurrence of a species in a particular area is dependent on percentage moisture within the soil. If the distribution of values of percentage moisture in the soil were quite random over the area, so that the value at any point were independent of the values at other nearby points, the distribution of individuals of the species would also be random, and no pattern due to its control by one factor would be apparent. In fact, as in this case, it is difficult to imagine most influencing factors having random distributions of levels, and the very widespread occurrence of non-randomness is in itself evidence that this normally is not so. The possibility must, however, be taken into consideration. Such a state of affairs can clearly occur in the special case of the influencing factor being the presence of an individual of another species, which may itself be randomly distributed. Thus, in the simplest example of a host-specific parasite, if the host is randomly distributed, the parasite will also be random unless the incidence is so high that more than one individual of the parasite may be dependent on a host individual. Such a clear-cut relationship between species is evident on inspection but less precise interrelationships occur, which are not so readily detected, e.g. the association between desert shrubs and annuals in California shown by Went (1942), and the effect of *Agropyron repens* on other species (Osvald, 1947; Hamilton and Buchholtz, 1955). These are attributable to such causes as the local altering of levels of moisture and mineral nutrients, critical to one species, around individuals of another species, or, in some cases, to the effect of root exudates.

When the factor influencing one species is the presence of individuals of another species, samples of suitable size will show association between the two species. This may be apparent either by correlation between quantitative measures of the two species or, if the effect is a more pronounced one, by the species occurring together in samples either more or less frequently than chance expectation, according as the influencing species favours or is antagonistic towards the influenced species. In the more general case of other influencing factors varying randomly over the area, indications of their importance may still appear from study of interspecific association. In any community of more than a few species it is unlikely that an influencing factor will influence one species only, and concurrent influence on several species will result in association between them.

Information of association between species is valuable apart from the detection of overriding influence on distribution where direct study of pattern fails. If several species in a community exhibit pattern, indicating control by influencing factors, study of association between them will provide evidence of any grouping of the species into assemblages of like response to the influencing factor or factors. Consideration of such species assemblages is similar to classification of plant communities. Indeed the distinction between species groupings in the present context of variation within an accepted plant community and species grouping into units accepted as different plant communities is largely one of magnitude of difference. Techniques of examination are similar and the distinction between aspects considered here and those postponed to the discussion of communities in Chapter 7 is rather arbitrary.

In assessment of association, as of pattern, provided that no measure of the degree of association is required, samples need not be random.* The null hypothesis tested is that there is no correlation between the occurrence of one species and another, and if this is true all samples are statistically independent of one another in this respect. The simplest procedure in the field is, generally, to use systematic sampling. There is a slight risk that equally spaced samples might give data from one phase only of a periodic variation in the vegetation. This risk can, in any case, be avoided by using contiguous quadrats, e.g. a grid or parallel lines of quadrats arranged in two sets perpendicular to one another. The contiguous quadrats are also advantageous in permitting grouping to give larger sizes. If a measure of degree of association is to be derived from the data, then the samples must be random, as otherwise no confidence limits can be assigned to the measure and no valid comparisons can be made between the measures for different pairs of species.

According as the species are present in all, or nearly all, samples

---

* There is one minor exception to this when the two species being considered differ widely in frequency (see p. 104).

or not, association between them may be tested in terms of correlation either of some quantitative measure of the species or of their presence or absence in the samples. There is rarely any choice of type of data to use for a particular case.

Presence or absence data for pairs of species can readily be tested in a 2×2 contingency table. Suppose the numbers of samples containing both species A and B, species A alone, species B alone, and neither species are as follows

Species A

	+	−	
+	a	b	$a + b$
−	c	d	$c + d$
	$a + c$	$b + d$	$a + b + c + d = n$

Species B

If the occurrence of species B is quite independent of that of species A, then of the $(a + c)$ samples containing species A a proportion $\dfrac{(a + b)}{n}$ would be expected to contain species B, i.e. the expected number of samples containing both species is $\dfrac{(a + b)(a + c)}{n}$. The expected number for the other cells of the table can similarly be calculated or obtained by subtraction from the marginal totals.* The observed numbers can be compared with this expectation based on independence by calculating χ^2 for the table, or by the exact solution, as appropriate (see Chapter 2, p. 39).

Unless interest centres on a few species only, it is generally worth while to make complete species lists for each sample, so that presence and absence data are available for a considerable number of pairs of species in a community. If a number of comparisons are made, the results of a few of them will individually correspond to a low probability by chance alone. Thus, if one hundred pairs of species are examined, the contingency tables for five of them will have an individual probability of 5 per cent or less and similarly one out of a hundred may be expected to show a probability of 1 per cent or less. No importance can therefore be attached to isolated

* This approach was apparently first used by Gleason (1925), who did not, however, make any test of significance of departure from expectation. More recently it was again suggested by Jones (1945), Cottam and Curtis (1948) and Cole (1949). Dice (1945) and others have emphasized the desirability of making such a contingency test before calculating any measure of association.

cases of apparent association when large numbers of comparisons are being made. This point has often been overlooked.

It is sometimes worth while to examine the joint occurrences of more than two species to clarify particular points. This may be illustrated from an example. Greig-Smith (1952b) found in a set of samples from secondary forest in Trinidad some evidence of association between three species, *Amaioua corymbosa*, *Lacistema aggregatum*, and *Alibertia acuminata*, which occurred in 58, 54, and 42 samples respectively out of 100. The observed and expected numbers of joint occurrences and probability levels of chance occurrence were:

	Observed	*Expected*	*P*
Amaioua–Lacistema	37	31·32	<0·05
Amaioua–Alibertia	29	24·36	0·05–0·1
Lacistema–Alibertia	29	22·68	<0·05

The suggestion that these three species formed an ecological grouping was supported by each showing negative association with two other species, themselves negatively associated. The expected numbers of joint occurrence of all three species in the various possible combinations are readily calculated, e.g. that for *Amaioua* with *Lacistema* but without *Alibertia* is $0·58 \times 0·54 \times (1—0·42) \times 100 = 18·17$. The observed and expected numbers in each of the eight possible classes are

	Am. Lac. Alib.	Am. Lac. —	Am. — Alib.	— Lac. Alib.	Am. — —	— Lac. —	— — Alib.	— — —
Observed	23	14	6	6	15	11	7	18
Expected	13·15	18·17	11·21	9·53	15·47	13·15	8·11	11·21
Deviation	+9·85	−4·17	−5·21	−3·53	−0·47	−2·15	−1·11	+6·79
χ^2	7·38	0·96	2·42	1·31	0·01	0·35	0·15	4·11

The total χ^2 is 16·69 with four degrees of freedom, corresponding to a probability of less than 1%. Not only is the departure from chance grouping of the species confirmed, with enhanced significance, but it becomes clear that the main contribution to the deviation is from samples containing either all three of the species or none of them. This makes more clear-cut the problem of the ecological significance of the association. This approach could be applied to the interactions of a number of species but it becomes impractical for more than three or four species as the number of possible groupings increases geometrically, e.g. there are sixty-four possible groupings of six species, and the expected number of occurrences in each group becomes low. In comparisons of joint occurrences of three or more species it may be necessary to group some classes so that no expected value is below 5, the usual arbitrary level for χ^2 calculations. There is not the same need to correct for continuity as, with the increased number of degrees of freedom, the number of possible tables for any set of 'marginal totals' becomes much greater and the probability distribution curve much more nearly continuous. The number of possible sets of species to be tested if they are taken in sets of more than two also increases rapidly with the number in a set, so that the labour involved in analysis becomes heavy. It is not generally worth while, therefore, to examine associations between more than two species unless there is strong indication from the 2×2 tables of interesting relationships involving more than two species. Alternative treatments are more appropriate to segregation of communities and will be considered in that context.

Comparison of observed with expected numbers of occurrences of species in samples is a very useful and flexible technique, which can be applied to a variety of situations. Where interest centres on certain categories of association only, records may not be available of samples containing no species. This situation may be illustrated from Went's (1942) data, already mentioned, on the association between shrubs and annuals in desert vegetation in California. Data were obtained along a belt transect of the number of shrubs of different species, the species of annual occurring round the base of each shrub and also of the number of occurrences of each annual species away from any shrub. There is thus no category of samples containing neither shrubs nor annuals; indeed there are no defined sample areas in the ordinary sense at all. The data are presented in the form of percentages but it is possible to reconstruct the data from the totals for different species. TABLE 6 shows the reconstructed data for four of the annual species, with the less numerous shrub species grouped together as 'other species'. Three principal questions may be asked: 1. Do the annuals under investigation tend to occur with shrubs, rather than away from them, more or less frequently than chance expectation? 2. Among the annual individuals found with shrubs is there evidence that they tend to occur more frequently with some shrub species than others? 3. If so, is the relationship the same for different annuals? In the absence of records of blank samples no direct answer can be given to the first question. An affirmative answer to the second, while certainly indicating association between annuals and individual species of shrubs, might arise from positive association with some species and negative with others. This would make the first question irrelevant, as the relationship with shrubs as a class would then depend on the relative numbers of different species. (In the present

TABLE 6

ASSOCIATION BETWEEN ANNUALS AND SHRUBS IN A DESERT IN SOUTHERN CALIFORNIA. OBSERVED (ABOVE) AND EXPECTED (BELOW) OCCURRENCES. DATA RECONSTRUCTED FROM WENT (1942)

		Encelia farinosa (living)	*Encelia farinosa* (dead)	*Franseria dumosa*	*Hymenoclea salsola*	Other species	Total	χ^2	P	No connection with any shrub
				SHRUBS						
	Number of shrubs in transect	248	84	172	136	133	773			
ANNUALS	*Phacelia distans*	74 112·93	49 38·25	88 78·32	92 61·93	49 60·56	352	34·45	<0·001	—
	Malacothrix californica	12 40·10	29 13·58	46 27·81	21 21·99	17 21·51	125	50·07	<0·001	—
	Emmenanthe penduliflora	10 31·12	30 10·54	6 21·58	27 17·07	24 16·69	97	70·49	<0·001	2
	Rafinesquia neomexicana	3 14·12	9 4·78	21 9·79	6 7·74	5 7·57	44	26·37	<0·001	1
	Total	99 198·27	117 67·16	161 137·51	146 108·73	95 106·33	618	104·69		

case a tendency for annuals to occur with shrubs was obvious on inspection and this justified the incomplete type of sampling used.)

If shrub species do not have differential effects on the occurrence of annuals, the number of occurrences of an annual species would be expected to be distributed among the different shrub species in proportion to the numbers of the latter. The expected numbers of occurrences calculated in this way are shown in TABLE 6 beneath each observed number. From each cell of the table an item can be calculated contributing to the total χ^2, which has 16 degrees of freedom. If the five items for each annual species are summed, a

χ^2 for each, having four degrees of freedom, is obtained. Each is highly significant, indicating that no annual species is randomly distributed among the shrub species. This could be due either to a general effect of the different shrub species on all annual species, or to specific effects between particular shrub and annual species. The total χ^2 for the table, 181·38, may be partitioned to test this. From the total observed and total expected number of occurrences of annuals for each shrub species (bottom line of the table) a χ^2 with four degrees of freedom can be obtained testing agreement of total annual occurrences with expectation. This is found to be 104·69. If this is subtracted from the total, the remaining portion is due to heterogeneity, or varying behaviour of different annual species in relation to the different shrub species. Thus we have

	χ^2	Degrees of freedom	P
Deviation	104·69	4	<0·001
Heterogeneity	76·69	12	<0·001
Total	181·38	16	

If there is a significant heterogeneity item, as in this case, it is important to notice that no importance can necessarily be attached to a significant χ^2 for deviation as, if the different annuals are associated in opposite senses with shrub species, the deviations of the total annual occurrences for different shrub species will depend on the relative numbers of different species of annuals. The analysis of χ^2 thus establishes that the annual species do not occur randomly with different shrub species, and that they differ in their interactions with different shrub species.

Analysis can often be carried further. Here, for instance, examination of the data suggests that a large part of the total deviation is due to annuals tending to occur with dead rather than living *Encelia farinosa*. All four annual species occur less frequently with living and more frequently with dead *Encelia* than chance expectation. Biological considerations suggest that dead shrubs are likely to be different in their effects on their immediate environment from those of living shrubs, regardless of species. In TABLE 7 the data for dead *Encelia* and for six* cases of occurrence with dead shrubs among 'other species' have been excluded and the marginal totals and expected values recalculated. The total χ^2 for each annual species is still highly significant as is the heterogeneity χ^2 for

* It has been assumed that the total number of dead shrubs is the minimum indicated by recorded cases of joint occurrence with annuals, i.e. the possibility of some dead shrubs having had no associated annuals is ignored.

the whole table, confirming that none of the four annuals is randomly distributed amongst the shrubs and that the former differ among themselves in their interaction with the shrubs.

TABLE 7

As Table 6, but Dead Shrubs Omitted

		SHRUBS				Total	χ^2	P
		Encelia farinosa	*Franseria dumosa*	*Hymenoclea salsola*	Other species			
	Number of shrubs in transect	248	172	136	127	683		
ANNUALS	*Phacelia distans*	74 110·02	88 76·30	92 60·33	49 56·34	303	31·14	<0·001
	Malacothrix californica	12 34·86	46 24·18	21 19·12	17 17·85	96	34·92	<0·001
	Emmenanthe penduliflora	10 22·88	6 15·86	27 12·55	20 11·71	63	35·90	<0·001
	Rafinesquia neomexicana	3 12·71	21 8·81	6 6·97	5 6·51	35	24·75	<0·001
	Total	99 180·46	161 125·16	146 98·96	91 92·41	497	69·42	

	χ^2	Degrees of freedom	P
Deviation	69·42	3	<0·001
Heterogeneity	57·29	9	<0·001
Total	126·71		

The data for *Malacothrix californica* and *Rafinesquia neomexicana* in Table 7 indicate similar responses to the different shrubs. If a similar table is drawn up with these two annuals only, the analysis of χ^2 is

	χ^2	Degrees of freedom	P
Deviation	57·62	3	<0·001
Heterogeneity	2·05	3	0·5–0·7
Total	59·67	6	

101

8—QPE

The heterogeneity χ^2 is here not significant, confirming that the interaction between these two species and the shrubs is, as far as the observations indicate, the same.

The examples quoted by no means exhaust the possible information obtainable from the data, but they are sufficient to show the flexibility of the contingency table approach.

If two species are present in all or nearly all the samples, association may be evident as a relationship between the abundance of the species. An obvious approach is to calculate a correlation coefficient between the abundance values for the two species. As an example, consider the following data for cover of *Cirsium acaule* and *Festuca ovina* in adjacent 10 ft. square plots on chalk grassland, each sampled by 112 points (seven frames of sixteen points). The figures are number of points out of 112 hitting the species, and have been arranged in order of cover of *F. ovina*.

Festuca ovina	*Cirsium acaule*
99	10
95	4
83	22
82	13
68	35
64	26
62	21
49	36
46	37

(With more extensive data the values for one species may be plotted against those for the other to see if there is any indication of relationship before calculating the correlation coefficient.)

There is some evidence that the cover of *Cirsium acaule* increases with decreasing cover of *Festuca ovina*, though the relationship is not very exact and cannot be accepted without testing. Denoting cover of *Festuca* by y and of *Cirsium* by x the necessary calculations are:

$(n = 9)$

$Sx = 204$, $Sx^2 = 5{,}776$, $\dfrac{(Sx)^2}{n} = 4{,}624$, $S(x - \bar{x})^2 = 5{,}776 - 4{,}624 = 1{,}152$

$Sy = 648$, $Sy^2 = 49{,}520$, $\dfrac{(Sy)^2}{n} = 46{,}656$, $S(y - \bar{y})^2 = 49{,}520 - 46{,}656 = 2{,}864$

$S(xy) = 13{,}074$, $\dfrac{SxSy}{n} = 14{,}688$, $S[(x - \bar{x})(y - \bar{y})] = 13{,}074 - 14{,}688 = -1{,}614$

102

Correlation coefficient

$$r = \frac{S[(x-\bar{x})(y-\bar{y})]}{\sqrt{[S(x-\bar{x})^2 S(y-\bar{y})^2]}} = \frac{-1614}{\sqrt{(1152 \times 2864)}} = -0.8886.$$

Variance of r $V_r = \dfrac{1-r^2}{n-2} = \dfrac{1-0.8886^2}{7} = 0.0300557.$

Standard error of r $\sqrt{0.0300557} = 0.1734.$

The value of t testing the departure of r from zero (the expected value if the values of x and y are independent) is $\dfrac{0.8886}{0.1734} = 5.13.$

This value of t has seven degrees of freedom, one having been used in determination of r, and the corresponding probability is less than 1 per cent.[*] The indication is thus confirmed that the cover of *Festuca ovina* is depressed where that of *Cirsium acaule* is greater, as indeed might be expected from the morphology of the latter.

Any quantitative measure may be used in tests of association and, indeed, different measures may be used for the two species. For example, if it were desired to test the relationship between an abundant grass species and a forb of low cover per individual, it would be appropriate to use cover for the grass and density for the forb. Hanson (1934) has described such a case where he found a (non-significant) correlation coefficient of -0.09 between density of stalks of *Agropyron Smithii* and cover of *Bouteloua gracilis*.

An alternative to the correlation coefficient is the use of a rank correlation coefficient, which has the advantages of greater ease of computation and of not giving undue weight to extreme values. Of those available the most suitable is Kendall's *tau*. (See, for example, Kendall, 1948.) To estimate this coefficient the data are arranged so that one variate is in natural order from the greatest to the smallest value and the order of ranks of the other variate is noted. The coefficient is then obtained by examining all possible pairs of values of the second variate, noting which are in natural and which in reverse order and expressing the excess as a proportion of the total comparisons. Thus, for the data discussed above, the rankings are:

Festuca ovina	1	2	3	4	5	6	7	8	9	
Cirsium acaule	8	9	5	7	3	4	6	2	1	
P	1	0	2	0	2	1	0	0	–	$+6$
Q	7	7	4	5	2	2	2	1	–	-30
S	-6	-7	-2	-5	0	-1	-2	-1		-24

[*] The values of r corresponding to certain probability levels for a range of values of n have been tabulated directly, e.g. Fisher and Yates (1943). The calculation of t is thus generally unnecessary.

The score, S, of excess of positive or negative comparisons, is most readily calculated in the way shown. For each entry for *Cirsium acaule* the number of entries to the right of it which are greater is noted as P and the number which are less as Q. S is equal to $P - Q$. The total number of comparisons made is $\frac{1}{2}n(n-1) = 36$. Thus $\tau = \frac{-24}{36} = -0.67$. It will be seen that for complete concordance in ranking of the two variates all comparisons will be positive and for completely reversed rankings all comparisons will be negative. Thus τ ranges from $+1$ for perfect positive association to -1 for perfect negative association.

For $n \geqslant 10$, S is approximately normally distributed and the significance of τ may be tested by comparing with its standard deviation and referring the ratio to the table of the normal distribution. The observed value of S should first be reduced by unity to correct for continuity. The variance of S is $\frac{1}{18}n(n-1)(2n+5)$ (Kendall, 1948). Kendall provides a table, which is reproduced in Appendix B (TABLE 8), for values of n up to 10. From this the probability in the present case is approximately 0·6 per cent.

If there are ties in the ranking of one or both species, the tied ranks are averaged, e.g. if the second and third highest values are equal they are both given the rank of $2\frac{1}{2}$. The number of possible comparisons is reduced and the coefficient then takes the form

$$\frac{S}{\sqrt{[(\frac{1}{2}n(n-1)-T)(\frac{1}{2}n(n-1)-U)]}}$$

Where T is $\frac{1}{2}\Sigma_t t(t-1)$, the sum of the expressions $t(t-1)$ for each group of t tied ranks in one variate, and U is, similarly, $\frac{1}{2}\Sigma_u u(u-1)$, the sum of the expressions $u(u-1)$ for each group of u tied ranks in the other variate. Under these conditions the variance of S is:

$$\frac{1}{18}\left\{ n(n-1)(2n+5) - \Sigma_t t(t-1)(2t+5) - \Sigma_u u(u-1)(2u+5) \right\}$$
$$+ \frac{1}{9n(n-1)(n-2)}\left\{ \Sigma_t t(t-1)(t-2) \right\}\left\{ \Sigma_u u(u-1)(u-2) \right\}$$
$$+ \frac{1}{2n(n-1)}\left\{ \Sigma_t t(t-1) \right\}\left\{ \Sigma_u u(u-1) \right\}$$

There remains one type of association to be considered, viz. where one species is present in all or nearly all the sample areas and the other is absent from many sample areas, and has a low measure in samples where it is present. Here interest centres on the relation between the amount present of one species and the presence or absence of the other. The only test possible is to classify the samples into those with and those without the second species and calculate the mean measure of the first species for the two classes. A t test may then be applied to the difference between the means. If this is to be done, sampling must be random, contrary to the general condition for tests of association, so that a valid estimate of the standard error of the two means may be obtained.

As in sampling for pattern determination, so in sampling to test association, care should be taken not to include obviously different communities in the same set of samples. If different communities are included, strong indication of association will be obtained, positive between species belonging to the same communities, negative between those belonging to different communities. It is true that these associations reflect the influence of controlling factors, but it is an influence that is apparent without quantitative examination, and the resulting associations may obscure less obvious relationships between the species. The effect of sampling an obviously heterogeneous area may be illustrated by an example. If two species have frequencies of 30 per cent and 40 per cent, and 14 joint occurrences (against random expectation of 12) are found in 100 samples, the corrected χ^2 is 0·45. If 100 samples containing neither species are added, the expected number of joint occurrences falls to 6, and the corrected χ^2 is 13·79 with probability of about 0·1 per cent. Bray (1956) has given a clear example of this effect in the field.

The effect of sample size on indications of association has been largely overlooked. Not only does it affect the indication obtained but useful information can be derived from the size of sample at which indication of association changes. There are four causes of association which may be distinguished by the behaviour of indications of association in relation to changing sample size.

1. If the quadrat used is of the same order of size as individuals, negative associations will appear which mean no more than that two individuals cannot occupy the same place. As quadrat size increases this association will disappear.

2. If some species present have individuals much larger than those of other species, spatial exclusion by the former may impose positive associations among the latter. Thus, if a species A has individuals large enough to exclude, from a quadrat of the size used, species B and C, having individuals small enough to occur together, these two species will show positive association. This effect is not uncommon, e.g. in grassland containing both tussocks of grasses, perhaps several

105

thousand square centimetres in area, and individuals of small forbs. Indication of association will disappear when the quadrat size is increased above the size of the individuals of the larger species.

3. If two species respond similarly to a controlling factor which has a defined pattern of values, they will show positive association up to the size of quadrat corresponding to the scale of heterogeneity of the controlling factor. With further increase in size of quadrat, the indications of association will disappear. Conversely, opposite responses to the controlling factor by two species will result in negative association, disappearing at a quadrat size above that of the scale of heterogeneity of the controlling factor. If several

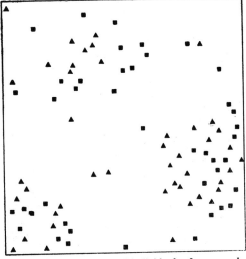

FIGURE 19. Distribution of individuals of two species
(see text)

controlling factors are operating, giving pattern with several scales of heterogeneity for the species, indications of association may change at two quadrat sizes. FIGURE 19 shows two species which are both much more abundant in certain parts of an area; within those parts the species are favoured by different levels of a second controlling factor, giving pattern on a second, smaller scale. Here, as quadrat size is increased, indications of association will at first be negative, then positive, and finally disappear. The positive association of two species imposed by spatial exclusion by a third, described above, is really a special case of this similar response to a controlling factor. That it is a simple spatial exclusion effect will generally be apparent on considering, in relation to the growth form of the species present, the quadrat size at which it disappears.

4. Direct effects between species, e.g. by root excretions or local modification of the soil factors round an individual, will, if causing positive association, give indications of association at all quadrat sizes large enough to include individuals of both species. If, on the other hand, the effect is one of negative association, indications of association will disappear when the quadrat becomes larger than the average area of influence of an individual. The result will be like that of spatial exclusion, but with indication of association not disappearing until a quadrat size greater than would be expected from simple spatial exclusion.

It is thus clear that the information derived from association data based on a single size of sample will be very incomplete and difficult to interpret. If circumstances allow, it is well worth while making observations with a number of sample sizes. This strongly reinforces the argument for using a scheme of systematic sampling, which will normally allow data for larger samples to be obtained by grouping adjacent samples. If circumstances permit only a single size, it may be advantageous with quantitative data to calculate partial correlation coefficients, taking account of the amount of other abundant species, as Dawson (1951) has suggested. This will eliminate spatial exclusion effects, but, as Goodall (1952a) has pointed out, may obscure relationships between groups of species.

One difficulty in interpretation of association data derived from a series of increasing sample sizes may arise from the change-over from presence or absence records to quantitative correlation. Comparison of quantities is the more sensitive test. Consider two species both more abundant in limited parts of an area only, and competing actively within those parts (cf. FIGURE 19). From the nature of the two tests it is almost always impossible validly to apply them both to the same set of data. Suppose, however, that examination of presence and absence at a sample size where both species have nearly 100 per cent frequency shows positive association, resulting from their common restriction to parts of the area. If the sampling unit is then increased slightly so that calculation of a correlation coefficient is the appropriate test, negative association may appear, owing to the strong competition effects between the two species where they are abundant being sufficient to outweigh their joint abundance in certain samples only. The negative association at the larger sample size represents a 'carry-over' of the spatial exclusion effect, made evident by the greater sensitivity of comparison of quantitative measures.

This reversal of indication at change to the correlation coefficient test may be illustrated from data given by Kershaw (1959). Pattern analysis of upland reseeded pasture by transects of 5 cm basic units (of five points for cover determination), had shown heterogeneity of *Agrostis tenuis*, *Lolium perenne*, and *Dactylis glomerata* at a block size of 64 units, determined by predominance of *Agrostis* on patches of shallow soil and of *Lolium* and *Dactylis* on deeper soil (see p. 91). Association tests of data from single units showed negative association, interpreted as spatial exclusion, for all three pairs of species. With blocks of 8 units *Agrostis–Lolium* and *Agrostis–Dactylis* showed negative and *Lolium–Dactylis* positive association, reflecting similar response by *Lolium* and *Dactylis* to the controlling factor. For blocks of 16 units it was necessary to change to calculation of a correlation coefficient. Most areas examined showed no association between *Lolium* and *Dactylis* but where their joint cover was high, i.e. where direct competition might be expected, there was evidence of negative association, though scarcely significant.

There has been considerable interest, especially among animal ecologists, in measures of degree of association as opposed to tests of association. Where a correlation coefficient can legitimately be used as a test its value can be used as a measure of association, provided the samples on which it is based are random, and that the correlated measures are normally distributed. Vegetation measures are frequently not normally distributed and if the correlation coefficient is to be used as a measure of association it will generally be necessary to transform the data first (see Chapter 2). Non-normality of the data does not affect the validity of the use of the correlation coefficient as a test of the existence of association. For data of the contingency type a wide range of measures has been proposed, many of which are unsatisfactory in some respect or other. The subject has been critically reviewed by Cole (1949) (see also Nash, 1950, Morisita, 1959b, and, for extension to coefficient of partial association, Cole, 1957). Only when the main interest is the relationship between the same pair of species in a number of different communities does it seem likely that a measure of association will be informative. Generally, the existence of association and the scale on which it appears are more important. Certainly the growing tendency to calculate a coefficient of association from presence and absence data and use its value as an indication of the existence of association has little to commend it; not only is the computational labour commonly greater than for a χ^2 test, but the validity of assessing the value of the coefficient by reference to its standard error is doubtful, especially when the total number of samples is small.

Pielou (1961) has introduced a concept of *segregation*, related to that of association, referring to the degree to which two species are intermingled, regardless of scales of pattern (FIGURE 20). This is

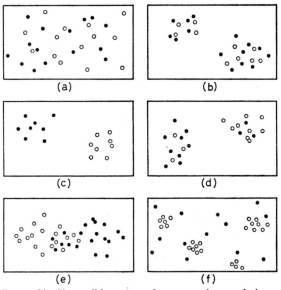

FIGURE 20. Six possible patterns for two-species populations: (a) and (b) unsegregated; (c) fully segregated; (d), (e) and (f) partly segregated (from Pielou, 1961, by courtesy of *J. Ecol.*)

approached in terms of nearest neighbour relationships. In a two species population four classes of nearest neighbour relationships may be distinguished:

(1) Individuals of A whose nearest neighbour is an individual of A
(2) . . . . . . A . . . . B
(3) . . . . . . B . . . . A
(4) . . . . . . B . . . . B

Denoting the numbers in these classes by f_{AA}, f_{AB}, f_{BA} and f_{BB} respectively a contingency table may be drawn up:

Base plant

		Species A	Species B	
Nearest	Species A	f_{AA}	f_{BA}	Na'
neighbour	Species B	f_{AB}	f_{BB}	Nb'
		Na	Nb	N

109

(N, total number of relationships examined; a, b and a', b', proportions of base plants and of nearest neighbours which are A and B respectively). This may be tested in the usual way by a χ^2 with one degree of freedom. Segregation will be indicated by the observed numbers of AB and BA relationships being less than random expectation. Pielou defines as a *coefficient of segregation*

$$S = 1 - \frac{\text{Observed number of AB and BA relationships}}{\text{Expected number of AB and BA relationships}}$$

$$= 1 - \frac{f_{AB} + f_{BA}}{N\,(a'b + ab')}$$

This can have negative values when AB and BA relationships are more numerous than random expectation. It ranges from -1 for isolated pairs made up of one A and one B to $+1$ for complete absence of AB and BA relationships.

The connection between segregation and association can be clarified by considering the cases illustrated and discussed:

Segregation	Indication of association with increasing quadrat size
Negatively segregated (isolated pairs AB)	positive at all sizes
Unsegregated FIGURE 20(a)	nil
FIGURE 20(b)	positive ⟶ nil
Fully segregated FIGURE 20(c)	negative ⟶ nil
Partly segregated FIGURE 20(d)	negative ⟶ nil
FIGURE 20(e)	negative ⟶ positive ⟶ nil
FIGURE 20(f)	negative ⟶ nil

Segregation is affected more by small scale effects than by large, just as indications of pattern obtained from plant-to-plant distances are more sensitive to small scale effects.

In a further consideration of segregation Pielou (1962b) has proposed fitting data for runs of different species in transects to various modifications of random expectation. This is analogous to the fitting of modified Poisson distributions to data for single species and is subject to the same objections (see p. 82).

Kershaw (1960, 1961) has considered the comparison of the patterns shown by different species in the same set of samples. The association between species at different block sizes may be examined in terms of presence or absence, or of correlation coefficient applied to the totals for the block size in question, as in the data quoted above. Kershaw has pointed out that the total covariance between two species may itself be analysed in terms of portions appropriate to different block sizes. If the representation of two species A and B in each unit of a grid or transect are added, an analysis of pattern of (A + B) may be prepared. If there is no association between the species, the resulting variance at any block size will equal the sum of the variances of the two species individually. If there is association then

$$V_{A+B} = V_A + V_B + 2C_{AB}$$

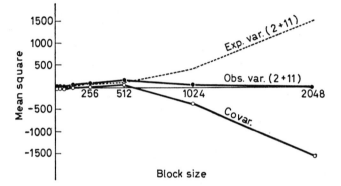

FIGURE 21. Expected and observed variance of combined data and covariance, plotted against block size, for two species. Data for cover along a transect extending from acidic into basic grassland. Species 2, *Festuca rubra*; species 11, *Holcus lanatus* (from Kershaw, 1961, by courtesy of *J. Ecol.*)

where C_{AB} represents the covariance at that block size. The covariance can then be plotted against block size (FIGURE 21). Since the correlation coefficient $r = C_{AB}/\sqrt{(V_A V_B)}$ it may readily be calculated and used in a test of significance. This approach has the advantage that the covariance isolated is that appropriate to the relationship of the species considered for blocks of x units within blocks of $2x$ units; the direct calculation of a correlation coefficient from the totals for blocks is liable to be affected by relationships at larger block sizes. It has the disadvantage of using product-moment correlation at the smallest block sizes, where there are likely to be numerous zero values and a contingency test is more appropriate.

CHAPTER 5

CORRELATION OF VEGETATION WITH HABITAT FACTORS

THE determination of gross differences in vegetation by environmental factors, as between the successive stages of a typical hydrosere, is accepted as self-evident; in such cases the factors or complex of factors responsible can generally be detected with reasonable certainty. On the other hand, it is uncertain how far the smallest differences between samples of vegetation are environmentally determined and how far they are matters of chance as to which species, of a number fitted to do so, happened to establish first. Intermediate degrees of difference between vegetation can commonly be related, with increasing certainty, to differences in environmental factors. At all levels, however, correlation between present environmental differences and present vegetation may break down owing to past changes in environment and vegetation, so that the determining factors are primarily historical ones. Compare, for instance, the tree growth resulting from secondary succession in areas completely cleared of forest at some previous time, with that occurring near enough to uncleared forest for the original species to return.

In so far as vegetational differences are determined by environmental differences, they may be expected to correlate with them. If the vegetational or environmental differences, or both, are small, then correlation will not be readily detected by qualitative examination, and only objective assessment of suitable quantitative data will reveal the relationship. This applies especially where several factors, themselves perhaps not independent, are jointly responsible. Evidence for determination by chance factors and, to some extent, by historical factors is negative. Their importance can only be assumed from apparent absence of determination by environmental factors. Important as the study of correlation between vegetation and the level of particular environmental factors thus is, it must be emphasized that mere correlation between the two variables is no proof of causal relationship either direct or indirect. Both may be determined by some other factor, e.g.

Conway (1938) points out that performance of *Cladium mariscus* is correlated with high temperature but the favouring factor is incidence of bright sunlight, which is responsible also for increased temperature. Additional information and general ecological knowledge may lend such support to a supposition of causal relationship that it is reasonable to accept a correlation as reflecting such relationship but final proof can generally be obtained only by experiment in which all other factors are held constant. Unfortunately, the appropriate experiment may not be possible because some other factors cannot be held constant, or may be impractically complex or lengthy to carry out.

Although most investigations have been concerned with the occurrence of a particular species or even of a definite state of a species (e.g. plants in fruit) in relation to environmental factors, exactly the same considerations apply to occurrence of vegetation types. Correlation of vegetation types with environment has rarely been considered in any detail, perhaps because of the difficulties of community delineation where small differences are involved. There is no theoretical reason why the distribution of communities should not be examined in relation to environmental factors in exactly the same way as species, though, normally, the abundance of a community cannot be stated in any precise way, and it must be recorded as present or absent only.

As in all quantitative ecological work, a suitable type of sampling* must be adopted if the maximum information is to be obtained on correlation between vegetation and environment. As sampling requirements vary to some extent according to the type of data being collected, sampling procedure will be considered in relation to the different categories of data discussed below. However, one misleading procedure, which has been used in some studies of the relationship between the distribution of species and such factors as soil acidity, must be mentioned first. Samples of the soil are taken from around the roots of a number of individuals of the species and the pH (or other factor) determined for each. If the number of occurrences at each pH value is plotted against the pH, a peaked curve, often approximately normal in form, but sometimes bimodal, commonly results. Some workers, e.g. Salisbury (1925), have interpreted the peak or peaks as optimal pH values. Volk (1931) and Emmett and Ashby (1934) have pointed out that the shape of the curves is affected by the frequency of soils of different pH values

* Sampling here refers to the positioning of samples. Environmental factors themselves should be determined in as efficient a manner as possible, but these aspects are outside our present scope and are, in any case, amply dealt with elsewhere.

available to the species, and that the frequency of different pH values around individuals of the species must be compared with the frequency of different pH values in the habitat examined, regardless of whether the species in question is growing there or not. As Ashby (1936) later pointed out, by such an erroneous procedure telegraph poles would show an 'optimal' pH value. Such data, drawn only from soil around the species, are adequate only to indicate the range of pH (or other factor) within which a species occurs and can provide no information on the behaviour in relation to varying values of the factor within the range tolerated.

Both plant and environmental data may be recorded either quantitatively or qualitatively. As in studies of association, the abundance and distribution of the species concerned will determine the form of the plant data. If the species is present in all, or nearly all, samples, then a quantitative measure will be necessary. If it is sparse, not only may recording merely of presence or absence be adequate to detect any relationship with the environmental factor, but quantitative measures may have so small a range of values, or so intractable a distribution curve, as to be impractical. Whether the environmental factor should be recorded quantitatively or not depends on the nature of the factor as well as the degree of precision desired. Some factors such as pH, light intensity and humidity can scarcely be classified other than by measurement. The degree of accuracy to which they are measured can of course be varied, and should be related to the magnitude of differences expected. On the other hand features such as soil texture, amount of litter, etc., can often be satisfactorily graded on inspection into a few categories, which may be sufficient to reveal relationships without the time-consuming physical or chemical analysis necessary for quantitative measures. Even such factors as the level of mineral nutrients may be graded as high, medium or low by approximate methods of analysis. Other factors, such as the presence and amount of undecomposed litter, where the contrast between presence and absence implies a much greater difference in environment than does a variation in amount of litter, can scarcely be treated other than qualitatively without confusing effects on two quite different scales. The form which the data for environmental factors should take thus requires careful consideration before the field work is begun.

Since both plant and environmental data may be either qualitative or quantitative, there are four categories of data to be considered:

1. Both plant data and environmental factors qualitative.
2. Plant data qualitative, environmental factors quantitative.

3. Plant data quantitative, environmental factors qualitative.
4. Both plant data and environmental factors quantitative.

The first category, with all records qualitative, is comparable to association data with both species recorded as present or absent only and the same considerations apply. It need not be considered in detail; it is sufficient to repeat that sampling may be either systematic or random, and that the data are examined in a contingency table. More than one environmental factor may be graded for each sample and the data treated in the same way as that for association between more than two species, but the results may be difficult to interpret. In general, examination of correlation jointly with a number of qualitatively recorded factors is not very profitable, though it may be useful where a species grows only where a number of rather narrowly defined environmental conditions are satisfied.

If the plant data are qualitative, and the environmental data quantitative, and a single environmental factor only is being considered, the test of relationship is straightforward. The mean values of the environmental factor for the plant classes (commonly 'with' and 'without' a species) are determined and compared by a t test. If several plant classes have been distinguished, a single comparison of their differences may be made in the form of an analysis of variance, comparing between-class differences with within-class differences, or the means may be compared in pairs, as seems appropriate. Data should not be combined in one analysis of variance if the variances for the different plant classes differ greatly from one another, as a significant difference may then be due to the differing variances rather than differing means of the classes. Widely differing variances may be allowed for in t tests in the way described earlier (p. 35).

Data of this type are recorded in two rather different circumstances, viz. 1, examination of the general difference in level of an environmental factor in two or more different communities spatially separate from one another; 2, investigation of the relationship between point-to-point variation in the environmental factor within a community and the distribution of individuals of species within the community. The two cases, though differing only in scale of differences involved, do present rather different practical problems of sampling. In the former case the several vegetation types must each be sampled randomly (with some restriction of randomization if deemed desirable) for estimates of the environmental factor. In the latter case those parts of the community with and without the species in question must likewise be sampled randomly.

In practice the two types 'with' and 'without' the species are often so intermingled that it is more convenient to allow each sample not only to be randomly selected in relation to the environmental factor, but also to be allotted randomly to the 'with' and 'without' classes. Samples taken at intervals along transects through the community have been used, e.g. Emmett and Ashby (1934) working on pH and occurrence of *Pteridium aquilinum* and *Vaccinium myrtillus*, Jowett and Scurfield (1949b) working on various soil factors and occurrence of *Holcus mollis* and *Deschampsia flexuosa*. Emmett and Ashby, working on large areas, drew arbitrary transects on a map and then followed them by compass, sampling at 50-yard intervals. Jowett and Scurfield, working in defined woodlands, drew transects selected 'at a fixed equal distance apart, the distance being based on the size of the woodland and selected arbitrarily beforehand as that giving the appropriate number of samples. At arbitrarily fixed equal intervals (also selected previously) along each transect, soil samples were taken. . . .' The latter procedure is in fact a completely systematic scheme of sampling, in which the position of the first sample determines that of the remainder, and hence is not a satisfactory basis for statistical analysis. Emmett and Ashby's procedure results in random sampling between transects (every point in the area has an equal chance of being represented) but the data from any one transect are linked in a systematic manner and, strictly, the data from one transect should be treated as one unit associated with 'presence' or 'absence' of the species. Clearly this cannot be done, so that this procedure is not entirely satisfactory either.

The transect method can be modified to give random samples. Equally-spaced transects are laid down in the way proposed by Jowett and Scurfield and a number of samples taken from each proportionate to its length. The actual position of each sample is fixed by two random numbers, one indicating distance along the transect and the other distance at right angles to it, the latter having a maximum value equal to the distance between transects. This results in a satisfactory form of restricted randomization, and one that is reasonably quick in the field. Alternatively, the common procedure may be adopted of using two lines at right angles as axes of random co-ordinates to give unrestricted randomization. It will generally be necessary to fix some arbitrary area around the sampling point within which the species is recorded as present or absent, e.g. Emmett and Ashby used a quadrat of area 10 square feet around the point at which the soil sample was taken. If the species being investigated is relatively sparse in the community, it may be necessary, in order to obtain sufficient data for the 'species present' class, to take the nearest individual to the random point selected and sample the environmental factor around or beneath it. This is unobjectionable, provided that no difference is made in the manner of taking the soil or other environmental sample from that used in the 'species absent' class, as, for example, by sampling a different soil layer.

Bieleski (1959) has described a case where it was desired to test the correlation with light intensity of tree seedlings of very low density. The usual approach would be to record the number of individuals in small quadrats together with the light value at the centres of the quadrats. The low density made this impracticable as a single light reading from a quadrat large enough to include a reasonable number of individuals would have little meaning. Bieleski therefore used a relatively large quadrat and recorded light intensity at each of a grid of points and also at all seedlings within the quadrat. The distribution of light intensities in the two sets of data were then compared. The sampling procedure is not entirely satisfactory as the data on the total range of light values are based on a systematic sample. Bias is unlikely to be serious as at the scale of the grid used (1 metre apart within 6 metre square quadrats in scrub) there is unlikely to be any correlation between points; random samples within the quadrat could equally well be used.

The testing of the observed difference in the values of environmental factors for different plant classes has given rise to some confusion. Bieleski has pointed out that difference in variance between 'species present' and 'species absent' classes is equally indicative of correlation between species occurrence and level of the environmental factor. Any difference in variance should therefore be tested before proceeding to a test of differences of means. A t test is, as we have seen, the appropriate test of difference of means in straightforward cases of two categories only, but the approach used by Emmett and Ashby (1934) and a criticism of it by Jowett and Scurfield (1949a) must be mentioned. As shown above, the sampling procedure used by Emmett and Ashby is not entirely satisfactory, but we may, for the present purpose, ignore this and consider only the handling of the data; data obtained by a satisfactory sampling procedure would have the same form. Their data are presented in the form shown in TABLE 8, which includes those for *Pteridium* and *Vaccinium* from one of the localities considered. They calculated the expected frequency of samples with and without *Pteridium* (or *Vaccinium*) in each pH class on the assumption of no correlation and, after grouping pH classes 4·8–5·1 and 5·8–6·1, tested the observed

117

frequencies against the expected by a χ^2 test. The expected frequency adopted for *Pteridium* is 51·772 which is the mean of the observed frequencies, omitting the first three classes, which contained no samples with *Pteridium*. This method of calculating the expected frequency gives a biased estimate, weighted in favour of pH classes containing a large number of samples. An unbiased estimate is obtained from the total number of samples, 212, and the number containing *Pteridium*, 108, and is 50·943 per cent. There is room for legitimate difference of opinion whether the results from the first three classes should be included in the calculation of expected

TABLE 8

FREQUENCY DISTRIBUTIONS OF OCCURRENCE OF *Pteridium aquilinum* AND *Vaccinium myrtillus* IN CLASSES OF pH (from Emmett and Ashby, 1934, by courtesy of *Ann. Bot., Lond.*)

pH class mean	Number of samples	Frequency of		Percentage occurrence	
		Pteridium	*Vaccinium*	*Pteridium*	*Vaccinium*
4·8	2	—	2	—	100
4·9	2	—	2	—	100
5·0	2	—	2	—	100
5·1	5	1	4	20·00	80·00
5·2	13	6	8	46·15	61·54
5·3	17	8	11	47·07	64·71
5·4	7	4	3	57·14	42·86
5·5	44	27	28	61·36	63·64
5·6	78	41	54	52·56	69·23
5·7	16	5	12	31·25	75·00
5·8	7	5	1	71·43	14·29
5·9	9	4	3	44·44	33·33
6·0	7	5	1	71·43	14·29
6·1	3	2	—	66·67	—

frequencies. It might be said that, since no *Pteridium* was found below pH 5·1, this represented the lower limit of tolerance. The number of samples with lower pH is, however, small and in this case it is perhaps wiser to attribute to chance the absence of *Pteridium* from them. If these classes are omitted in calculating expected frequency, the expected frequency should not be applied to them nor should their deviations be included in the χ^2 as Emmett and Ashby did. There is also an error, as Jowett and Scurfield pointed out, in calculation of the χ^2 in that the contribution of the deviations in the *Pteridium* absent class is omitted.

If the primary intention in making these observations was, as Emmett and Ashby imply, to test whether pH was correlated with the presence or absence of a single species, the appropriate test is comparison of mean pH for samples with and without that species. If this is done the values of t are 2·02 for *Pteridium* (probability slightly less than 5%) and 3·40 for *Vaccinium* (probability *ca.* 0·1%), indicating that there is some correlation in this habitat between pH and the occurrence of the two species, contrary to Emmett and Ashby's conclusions.* If a χ^2 test is used, based on an expected occurrence of *Pteridium* of 50·943 per cent in each pH class, a total χ^2 of 13·685 with seven degrees of freedom is obtained, corresponding to a probability of between 5–10 per cent. A substantially similar conclusion is drawn but the χ^2 test is evidently less sensitive, owing to grouping of the extreme classes necessary to bring all expectations above 5. Jowett and Scurfield apparently assume that the object of the observations was to determine whether *Pteridium* and *Vaccinium* had different pH 'preferences'. They compared the mean pH for samples containing *Pteridium* but not *Vaccinium*, with that for samples containing *Vaccinium* but not *Pteridium*, arguing that if neither species is influenced by pH, the two categories should show no difference in mean pH; the difference corresponded to a t of 2·85 (probability 2%). This method of testing provides information of interest but it brings into consideration not only the influence of pH on the occurrence of *Pteridium* but also the influence of the presence of *Vaccinium* on occurrence of *Pteridium* and *vice versa*. The situation may occur in which two species separately are not correlated with pH, but their relative competitive strength is. In this case significant difference would be shown by Jowett and Scurfield's test. This test is further open to objection that part of the data, those samples containing both species or neither, is ignored.

If it is desired to consider the correlation of the occurrence of the two species with one another as well as with pH, this may be done by an analysis of variance including the interaction between the species. TABLE 9 shows this analysis for Emmett and Ashby's data.† It will be seen that both the primary differences for presence and absence of *Pteridium*, and of *Vaccinium*, and the interaction between them, are significant. Too much reliance cannot be placed on the

* If, however, the first three pH classes are omitted in the *Pteridium* calculations, providing a more stringent test, t is 0·957, with probability 30–40 per cent.

† Note that here, as is likely to be the case with most data of this type, the numbers of samples in different subclasses are not proportionate and the preparation of an analysis of variance is not so straightforward as for the usual orthogonal data. See, for example, Snedecor (1946, § 11.10).

119

conclusions in this case, owing to the unsatisfactory features in the original sampling already discussed. If the data were accepted they would indicate not only definite pH 'preferences' by the two species but also differential effects on one another at different pH levels. There is, however, no direct information from the analysis as to whether the mean pH value for samples with *Pteridium* differs from that for samples with *Vaccinium*. It could be obtained by comparing the mean pH of all samples containing *Pteridium* with that of all samples containing *Vaccinium*, but would have little ecological meaning in view of the interaction between the species in relation to pH.

This contingency approach used by Emmett and Ashby may sometimes be useful. If the environmental factor has been measured with such a low degree of accuracy relative to the range observed

TABLE 9

ANALYSIS OF VARIANCE OF pH IN SAMPLES WITH AND WITHOUT *Pteridium aquilinum* AND *Vaccinium myrtillus* (data of TABLE 8)

	Degrees of freedom	Mean square	F	P
Pteridium present v. *Pteridium* absent	1	0·36136	7·79	0·001–0·01
Vaccinium present v. *Vaccinium* absent	1	1·00564	21·68	<0·001
Interaction	1	0·65100	14·03	<0·001
Error	208	0·04386		

that the number of different observed values is small, and this sometimes applies to pH, the distribution is so discontinuous that it is better to regard the values as grades of the environmental factor and test by contingency χ^2. Another difficult situation may arise if the distribution curve of the environmental factor in all samples together departs widely from the normal, and cannot be brought even approximately to normal form by transformation, so that the *t* test is inappropriate. This would apply if the curve of pH values were markedly bimodal, indicating perhaps that two quite different environments had been included in the samples. Here again the contingency approach is appropriate, though it would be better to repeat the observations, and to separate the supposedly different environments, if they can be identified. This may not always be possible, and the non-normal distribution curve may sometimes prove to be a real feature of a single environment. Finally, it may be noted that the contingency approach will take account of difference of variance as well as difference of mean.

It is sometimes useful to take several measurements of the environmental factor at each sampling site. This applies especially to soil characteristics which may vary up and down the soil profile. Such data are conveniently compared by an analysis of variance classifying the values for the environmental factor not only according to the plant class but also by position in the soil profile. This may be illustrated by the data in TABLE 10, showing values for percentage calcium carbonate at the surface and depths of 6 and 12 inches in five different vegetation zones from the foreshore community (stage A) to fixed dune pasture (stage E) (Gimingham *et al.*, 1948). Replication of samples was provided by laying down three transects at right angles to the vegetation zones and taking one sample at random within each vegetation zone on each transect. The regular pattern of the vegetation zones as strips parallel to the shore permitted this linking together of samples from different stages as units. Where, as more commonly happens, the vegetation pattern is irregular, replication would take the usual form of random sampling within each vegetation type and a transect item would not appear in the analysis of variance. Observations had suggested that different positions along the shore received differing amounts of wind-blown sand. This prompted the decision to remove an item in the analysis representing the sum of squares for transects, which decision was thus an ecological rather than a statistical one. The highly significant mean square for transects confirmed the correctness of doing so. We are not concerned here with the detailed conclusions to be drawn from the data; broadly they indicate little difference in total carbonate in the different vegetation stages (stages not significant), different concentrations at different depths (depths significant), and varying relation between carbonate and depth in different stages (interaction significant), probably owing to differing balance between leaching and accretion of sand.

There remains the case of concurrent measurements of several environmental factors, with qualitative plant recordings. Each environmental factor may be tested in turn for correlation with species occurrence, but it is clearly desirable to compare the total environmental information with the presence or absence of a species or vegetation type. General ecological experience suggests that sometimes an extreme value of one environmental factor, which would otherwise be unfavourable to the survival of a species or type of vegetation, may be compensated by an unusually high or low value of another factor, itself perhaps detrimental in other conditions. (Compare, for example, the occurrence of *Calluna vulgaris*, normally calcifuge, on calcareous soils under extreme

TABLE 10

PERCENTAGE CALCIUM CARBONATE IN DUNE SAND
(from Gimingham *et al.*, 1948, by courtesy of *Bot. Soc. Edinb.*)

Vegetation	Depth	Transect			Mean	Stage mean
		I	II	III		
Stage A	Surface	68	57	56	60·3	
	6 in.	68	70	61	66·3	60·7
	12 in.	60	52	54	55·3	
Stage B	Surface	59	50	51	53·3	
	6 in.	62	56	52	56·7	57·7
	12 in.	67	67	55	63·0	
Stage C	Surface	54	50	57	53·7	
	6 in.	60	69	58	62·3	59·1
	12 in.	69	64	51	61·3	
Stage D	Surface	64	62	57	61·0	
	6 in.	62	53	58	57·7	58·8
	12 in.	59	53	61	57·7	
Stage E	Surface	52	48	48	49·3	
	6 in.	68	62	65	65·0	55·8
	12 in.	61	53	45	53·0	

Depth mean

Surface   55·6
6 in.   61·6
12 in.   58·1

ANALYSIS OF VARIANCE

	Sum of squares	Degrees of freedom	Mean square	F	P
Stages	118·80	4	29·700	1·49	> 0·2
Depths	278·53	2	139·267	7·01	0·001–0·01
Interaction	608·13	8	76·017	3·82	0·001–0·01
Transects	370·53	2	185·267	9·32	< 0·001
Error	516·80	26*	19·877		
Total	1,892·80	42*			

* Total and error degrees of freedom are reduced by two because two samples were lost and their values estimated. (See Snedecor, 1946, § 11.6 for discussion of estimation of missing data.)

climatic conditions.) Hughes and Lindley (1955) have pointed out the value of the D^2 statistic developed by Mahalanobis and by Rao (see Rao, 1952) for this kind of data. D^2 is calculated from the means, variances and covariances of the various factors measured in replicate samples of two or more different groups. If only one factor is measured D^2 reduces to the t test. Like t it represents a measure, in terms of the factors measured, of the amount of difference between groups of samples. The probability of getting a given D^2 by chance can be determined, thus giving an indication of the significance of difference between groups. Its use may be illustrated from an example quoted by Hughes and Lindley. Determination of exchangeable calcium, available phosphate, exchangeable potash

TABLE 11

D^2 ANALYSIS OF SNOWDONIAN SOIL SERIES (from Hughes and Lindley, 1955, by courtesy of *Nature, Lond.*). FIGURES ARE THE VALUES OF D^2 FOR THE COMPARISONS INDICATED. (SEE FIGURE 22 FOR KEY TO SOIL TYPES)

	A′	B	B′	C	C′
A	2·66***	3·66***	4·46***	0·26	3·07***
A′		1·91***	0·30	2·18***	1·44***
B			2·44***	2·21***	2·48***
B′				3·76***	1·78*
C					2·09**

Probability   * 0·01–0·05
         ** 0·001–0·01
         *** <0·001

and pH were available for a number of samples of each of six soil types in Snowdonia. D^2 was calculated for each pair of soil types, giving the values shown in TABLE 11. These indicate that the soil types A′ and B′ are not significantly different from one another in terms of the four properties measured, nor are the soil types A and C. The six soil types may thus be grouped into four classes, which are shown in FIGURE 22 with the lines between them proportional in each case to the amount of difference between the groups. This example refers to sets of samples defined by the soil type but they might equally well have been defined by the vegetation they bore. The application of this technique has evident possibilities in environmental studies. Though the computational labour is rather heavy the value of being able to integrate data for different environmental factors is likely to make it worth while, at least in intensive studies.

The third category of data, with qualitative information on the environmental factor and quantitative plant data, can be dealt with more briefly, as considerations in handling the data are essentially similar to those in the last category. The basic technique is the comparison of means of the vegetation measures for the different environmental classes, either by t test for pairs of environmental classes or by analysis of variance for a number of classes considered together. More complex situations may be dealt with by methods analogous to those discussed in connection with the last category. When the environmental classes are spatially well

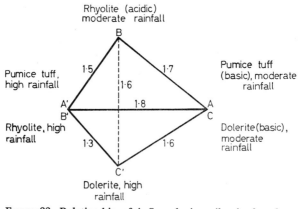

FIGURE 22. Relationships of six Snowdonian soil series, based on D^2 analyses. Figures are the values of D between series (from Hughes and Lindley, 1955, by courtesy of *Nature, Lond.*)

distinct from one another the problem resolves itself into comparison of a vegetation feature in two distinct communities, i.e. comparison of overall composition, already discussed in Chapter 2. If a number of species are measured in the same series of observations, and it is desired to consider their relationships together, the problem differs only in scale from that of distinction and delimitation of communities, which will be discussed in Chapter 7. The requirements of sampling will be apparent by analogy with those for the last category. Samples may be selected for the environmental factor, but within the selected environments the vegetation must be sampled randomly. It may, however, be more convenient to allow a sample to be randomly assigned to environmental classes if the latter are closely intermingled.

The last category of data includes those cases with quantitative measures of both plant and environmental factor. The sampling procedure is straightforward. Since the null hypothesis is that the two measures are unrelated, there is no need for either to be sampled randomly. Practical considerations will normally dictate the generally quicker and easier method of systematic sampling. The data obtained bear an evident resemblance to data on association between species when both species are recorded quantitatively, where, as we have seen, the appropriate test is the calculation of a correlation coefficient. In this case, however, there is an important difference in the nature of the data. When testing association between species it can be assumed initially, in the absence of any indications to the contrary, that the part played by the two species in the relationship is similar. The results of association tests may be used to postulate influence on one species by another, but generally the investigator is concerned to make a general test only of interrelationship in occurrence and, if any relationship is suspected, it is commonly control of both species by an environmental factor that is in mind. In testing correlation of vegetation with an environmental factor, however, the two variables can normally be assumed to have different rôles in the relationship. In the absence of evidence to the contrary it is reasonable to assume that the level of the environmental factor is determining directly or indirectly the level of abundance of the plant. It may be that both are controlled by some other environmental factor, in which case the correlation coefficient is perhaps more appropriate, or even that the environmental factor is controlled by the level of the species, as might apply, for example, to atmospheric humidity, but the initial assumption of the controlling part played by the environmental factor is nearly always justifiable ecologically. In these circumstances the use of regression analysis is more appropriate.

The reason for preferring regression analysis to the correlation coefficient, which is concerned only with the degree of interdependence, will be clearer if the derivation of a regression line is considered. If concomitant observations are made on any two variables, e.g. plant density and soil moisture content, the results may be plotted on a graph with the y and x axes representing plant density and soil moisture content respectively. If there is a relationship between the two, the points will not be scattered over the whole area of the graph, but will approximate more or less closely to a line, according to the exactness of the relationship. If the fit to a line is reasonably good, the appropriate line may be sufficiently clear on inspection. The fitting of a regression line is the calculation of a line (straight or of other defined form) which best fits the data. If we calculate the linear regression of y (the *dependent variate*) upon x (the *independent variate*), the line fitted is $y = a + bx$, in which the constants a and b are so fixed that the sum of squares of deviations of all points from the line, measured in terms of y, are minimized.* Conversely, the regression line for x upon y, generally written as $x = a + by$, minimizes the deviations measured in terms of x (FIGURE 23). Clearly the two lines will not necessarily be the same. In one case the fitting depends primarily on values of y, in the other primarily on values of x. Put another way, the regression of y upon x measures the expected change in y for a given change in x, and the regression of x upon y the expected change in x for a given change in y. In the example quoted, if the soil moisture level is

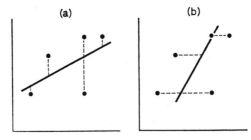

(a)　　　　　　　　(b)

FIGURE 23. Regression lines of (a) y upon x and (b) x upon y for the same points. The broken lines show the deviations minimized by the regressions

controlling the density of the plant, it is logical and justifiable to estimate the increase or decrease in density for a given change in soil moisture content, but the reverse process is ecologically meaningless. In the present context we are concerned less to evaluate a relationship than to show that it exists. The existence of a relationship is shown by the line making an angle with the axes, i.e. by the coefficient b having a value differing significantly from zero. Like any other statistic calculated from data, b is subject to random variability and appropriate tests are available to determine the probability of chance occurrence of a value of b as great as that actually obtained, and hence of deciding the significance of an indicated relationship.

We have so far considered the relationship only in terms of a straight line. Within the limits of a particular set of observations

---

* The line is actually calculated as $y = a' + b(x - \bar{x})$.
$$a' = \bar{y}, b = \frac{S(x - \bar{x})(y - \bar{y})}{S(x - \bar{x})^2}$$

this may approximate to the truth. Equally, however, the relationship may be expressed on the graph by a curve. It may be perfectly possible to derive an equation connecting x and y in some other way, e.g. $y = a + bx^2$, $\log y = a + bx$, $\log y = a + b \log x$, etc. The likelihood of such a line fitting the data should be judged, before embarking on the calculations, by plotting the transformed values of x and/or y and seeing if the relationship then approximates to a straight line. If no straight line relationship is obtainable in this way, a satisfactory fit can often be obtained to an equation with several terms in x, e.g. $y = a + bx + cx^2 + dx^3$, etc. To do so, however, from the point of view of establishing a relationship with environmental factors, is a dubious procedure, as there is usually no possibility of placing a biological interpretation on such a complex relationship. Before trying any such relationship careful thought should be given to the meaning of the type of equation being tested.* That is not to say that such relationships should never be considered. If, for example, a measure of abundance was being correlated with altitude, the relationship might really be with various meteorological factors of which some varied directly as altitude and others as the square root of altitude, in which case a relationship of the type $y = a + bx + c\sqrt{x}$ would be suitable.

If several environmental factors have been measured, the joint relationship of the plant measure with them cannot be plotted as a simple two-dimensional graph, but the same type of treatment may be applied to the data, e.g. if two environmental factors, represented by x_1 and x_2, have been measured, then a *multiple regression* $y = a + bx_1 + cx_2$ may be calculated. The significance of the whole regression and of the coefficients of each independent variate may be tested. The mode of calculation takes into account any correlation between the independent variates.

Little remains to be said on the treatment of data of the fully quantitative type. It may, however, be helpful to quote an example of the type of relationship that may be found. Rutter (1955) made an extensive analysis of the relationship between percentage contribution to the total foliage of *Molinia caerulea*, *Erica tetralix*, and *Calluna vulgaris* in an area of wet heath vegetation and various factors concerned with water relations. The percentage contribution to the foliage, a novel measure well suited to three species of rather diverse morphology, was estimated by eye, preliminary observations having shown a high degree of correlation between such estimates

* It must also be remembered that perfect fit to a set of n observations can always be obtained by an expression with $n - 1$ terms in x.

and actual separation and weighing of the foliage. The following are the relationships found for one set of data:

Molinia caerulea	$y = 24\cdot4 + 0\cdot78s + 1\cdot91t$	
Calluna vulgaris	$y = 29\cdot0 + 0\cdot16s - 1\cdot21t$	
Erica tetralix	$y = 50\cdot3 - 0\cdot94s - 1\cdot75t$	

where y = angular transformation of percentage contribution to the foliage, s = mean depth of summer water table below the surface, t = mean height of *Molinia* tussocks (a factor found to affect not only the performance of *Molinia*, but that of other species also). All the coefficients are highly significant, except that of s for *Calluna*, which is not significant.

There is no limit to the number of independent variates that may be included in a multiple regression. For example, Medin (1960) examined the regression of yield of the browse shrub *Cercocarpus montanus* on twelve soil site factors. The labour of computation becomes very heavy as the number of variates increases, but the increasing availability of electronic computing facilities is removing this difficulty. Thus Fritts (1960), using electronic computation, has examined the relationship between daily radial growth of *Fagus grandifolia* and nineteen independent variates, mostly climatic. The relative ease of computation does not, however, remove the need for careful prior assessment of the relevance of the variates to be included; unnecessary elaboration of the field and observational programme is as wasteful as is unnecessary computation.

Sometimes one environmental factor may be recorded qualitatively and others quantitatively. The former may be included in an analysis of variance in a neat and satisfactory way used by Blackman and Rutter (1946). One of a number of sets of observations on the relationship between density of *Endymion non-scripta* and light intensity was in a mixed plantation of oak and larch. It was suspected that the nature of the canopy, oak or larch, might be affecting the density of *Endymion*, apart from the differing light intensities beneath the two species. They included in the regression a 'pseudovariate' taking the value 1 if a sample was beneath oak canopy and zero if it was not, and obtained the multiple regression $y = 35\cdot11x - 3\cdot33z - 4\cdot26$, where y = square root of density, x = light intensity as fraction of full daylight, z = 1 if beneath oak canopy, 0 if not. From this regression equation it follows that at a given light intensity the density of *Endymion* is less beneath oak canopy than beneath larch.

Hughes (1961) has suggested the application of canonical correlation to ecological data. This involves the derivation of variates of the form $U = a_1x_1 + a_2x_2 + \ldots$, $V = b_1y_1 + b_2y_2 + \ldots$ in which the constants a_1, a_2, ... and b_1, b_2, ... are such that U is the most efficient linear combination to predict V and *vice versa*. This approach has evident possibilities in the study of the joint occurrence of several species in relation to the levels of several environmental factors, particularly if it is suspected that two or more species are closely similar in their requirements. It has apparently not been tested on field data.

There is one exception to the use of systematic sampling for the study of relationships with quantitative data. Sometimes interest centres on the nature of relationships rather than on their existence, e.g. information may be wanted whether the difference in abundance of a species for a given difference in an environmental factor is the same in two different communities. This is normally obtained by comparing the regression coefficients obtained from the data from the two communities. If the sampling is systematic there is no means of testing the difference between the coefficients as no unbiased estimate of their variances is available. Just as in the comparison of means the sampling on which they are based must be random if their variances and so the variance of the difference between them is to be estimated, so, if two regression coefficients are to be compared, the values of the dependent variates must be obtained from random samples. This should be borne in mind in planning observations if the need to compare coefficients with one another is likely to arise.

No mention has been made of the number of samples necessary in observations on vegetation and environmental factors. No single guiding principle can be given. For data of the first category it is necessary that no expected class of observations should be so low that the most extreme deviation has a probability greater than that adopted as the limit for significance. If a mean is being determined, as in the second and third categories, the approximate methods of deciding sample size discussed in Chapter 2 may be applied, but the number of samples necessary for a stringent test clearly depends on the magnitude of the observed difference. It may be helpful to plot successive values of the difference of means, rather than the means, and continue sampling until the difference has reached a more or less steady value. For the fourth category little guidance is available except previous experience with similar observations, particularly if multiple regressions are involved. As such observations are generally time-consuming it is well not to collect too extensive data until a preliminary examination of a relatively small number of observations has shown whether the approach is a promising one.

The four categories differ in their convenience of sampling and analysis and, to some extent, in their sensitivity. There is, however, rarely much choice of category for a particular investigation. Whether plant data are treated quantitatively or qualitatively depends on the number of blank entries and the range of values recorded. Environmental data likewise can generally be recorded only in the one form or the other, unless an arbitrary boundary in quantitative data is drawn between 'high' and 'low' values.

CHAPTER 6

PLANT COMMUNITIES—I. DESCRIPTION AND COMPARISON

IN a sense the preceding chapters have all dealt with plants as they occur in communities. We have been concerned with the information on abundance and spatial distribution that may be obtained from samples taken within a defined area, the vegetation of which may, in the broadest sense, be termed a plant community. The emphasis has, however, been more on the differential behaviour of individual species both within and between such communities. In this chapter and the next we are concerned with the characterization and comparison of the whole assemblages of species constituting the communities. Consideration of these leads on to problems of classification of communities, whether any natural classification is possible and, if so, how it may be achieved. This is fundamentally a separate subject, but it is so bound up with characterization of communities that the two can scarcely be considered apart.

Any discussion of plant communities as units raises the question whether or not they are more than abstractions made by ecologists from vegetation, the variation of which is continuous in space and, perhaps, in time. On the one hand there is the view that there are no discontinuities in natural vegetation, except where there are discontinuities in the physical environment, as at the junction between the outcrops of different geological strata. This view, of a continuum of vegetation, implicit in the outlook of many ecologists, was clearly stated by Gleason (1926) and has, more recently, been adopted and developed especially by Matuszkiewicz (1948), by Curtis and his associates (see especially Curtis & McIntosh, 1951; Brown & Curtis, 1952; Curtis, 1959) and by Whittaker (1952, 1956, 1960). At the other extreme there is the view of the community as an organism (Clements, 1916) or quasi-organism (Tansley, 1920), postulating that the individuals and species within a community so interact as to increase one another's potentiality of survival. This concept necessarily implies more or less sharply defined boundaries between one community and another. Conversely, if communities

131

do have sharply defined boundaries it follows that they must have some degree of reality as units.

Rejection of the organismal concept of the community, it may be noted, does not necessarily exclude the existence of separate communities as units, quite apart from the effects of environmental discontinuities. If species had ranges of tolerance in relation to environmental differences that tended to coincide, so that the total number of species in a region could be arranged in a considerably smaller number of groups, the members of each having approximately the same limits of tolerance, then distinctive communities, with more or less well-defined boundaries, would be expected, each corresponding to, and composed of, one of the groups of species of similar tolerance. The existence of such a phenomenon seems to be implied in the approach of both the Zürich–Montpellier and Uppsala schools of plant sociology, both of which erect a hierarchical system of classification of communities, without accepting any organismal or quasi-organismal view of the community. No extensive examination of the limits of tolerance of a geographical group of species in relation to all environmental factors has apparently been made; indeed, it is probably impossible except as a theoretical ideal. The work of Curtis and his associates, quoted above, points against any general coincidence of limits of tolerance as does most work on individual environmental factors, e.g. Hora (1947) on pH.*

Reference has been made to systems of classification because their erection carries implications of the existence of discrete communities on the ground. A clear distinction must be made between the abstract concept of vegetational units of different grades, embodied in a system of classification, and the actual existence in the field of communities with defined boundaries between them. The real existence of the classificatory units implies the real existence of communities in the field. The converse is not necessarily true; communities might be clearly distinguishable in the field, but the communities so distinguished might form a continuum not susceptible of classification except by drawing arbitrary boundaries. The distinction between the units of classification and communities is analogous to that between species and individual plants.

---

* The fact that classification of communities by the methods used by the Zürich–Montpellier school is possible might be considered evidence in favour of coincidence of limits of tolerance, but the successful results are invalidated as evidence by the insistence on floristic homogeneity in the stands selected for description. The results of the Uppsala school are irrelevant here as the classification is subjective. Cf. Goodall (1952a) on the common misconception of the Uppsala classification as based on objective discrimination.

Goodall (1954a) has attempted an objective assessment of the reality of plant communities. He argues that a community, if it has real existence, should show homogeneity of composition within its boundaries and cites evidence that homogeneity is not found on any of three criteria, (*a*) spatial distribution of individuals, i.e. agreement with Poisson expectation, (*b*) lack of correlation between quantities of species in replicate samples, (*c*) constancy of variance with increasing quadrat spacing. He rightly suggests that, although complete homogeneity may not exist, there may be a much greater degree of homogeneity within one stand (community in the sense used here) than between different stands. This difference in level of homogeneity may be sufficient to allow delineation of stands. It follows from this that, if samples are drawn from more than one stand, of pairs of samples those near together are more likely to be drawn from the same stand than those far apart. Hence variance between adjacent samples should be less than between those farther apart. He quotes data showing that in several areas examined this is not so. Unfortunately, in each case the samples appear to have been taken entirely within an area the vegetation of which would be regarded on most subjective criteria as a single community, and the investigations refer rather to pattern of the type discussed in Chapter 3 than to the differences between communities. The approach is an interesting and original one, but needs to be applied to areas including more diverse vegetation before it can assist greatly in assessing the reality of communities.

Transects which include a transition between clearly distinct types of vegetation commonly illustrate the difficulties of delineating the boundaries of communities. If the occurrence of species along such a transect is noted, it is rarely found that the points at which they appear or disappear from the record coincide even approximately, unless there is a very pronounced environmental discontinuity. There is in fact great difficulty in drawing any boundary between the two types, although there may be a length of the transect in which change in composition with position is very much more rapid than elsewhere, i.e. there is a more or less well-defined zone of transition or ecotone. The question has been well summarized by Webb (1954). 'The fact is that the pattern of variation shown by the distribution of species among quadrats of the earth's surface chosen at random hovers in a tantalizing manner between the continuous and the discontinuous'.

The reality or otherwise of distinctive units of vegetation in the field remains to some extent an open question and one likely to

133

remain a matter of considerable interest for its bearing on differing concepts of the plant community. It has, however, relatively little importance in many fields of ecological interest. Empirical description of vegetation cannot wait for clarification of theoretical concepts and ecologists are well aware that satisfactory approximate boundaries can often be drawn between different vegetation types. Even if no approximate boundaries can be fixed, it is still possible to delimit arbitrarily areas away from doubtful transition zones, areas which may be studied in as exact a manner as possible. Such a procedure is quite justifiable provided it is remembered that the information obtained applies only to the area delimited, and that when it is compared with that from another area and found different, the two areas may represent the end points of a continuous series the intermediates of which have been rejected as transitional. Though theoretically this is a serious limitation, for many practical purposes the disadvantage is lessened because the very fact of an area being recognized subjectively as distinctive and not transitional is evidence that it represents a significant proportion of the total vegetation of a region.

However the boundaries of areas selected for description are fixed, the technique of characterization and comparison of the vegetation of areas is the same. Moreover, they apply whether the objectives of investigations are contributions to ecological theory, empirical cataloguing of vegetation, or information of immediate economic importance in agriculture or forestry. We now turn to consideration of these techniques. It will be convenient in doing so to use the term 'stand' to denote any area the vegetation of which has been treated as a unit for purposes of description. This will help to emphasize that we are dealing with concrete samples of vegetation and not with abstractions of classification.

No two stands can be identical, if sufficient detail is taken into account. Moreover, they may differ in so many characters that it is impossible to take all of these into consideration. A subjective selection must, therefore, be made of criteria to be used in characterization and comparison of stands. This selection will be made in the light of the kind of information desired and the scale of differences to be examined. Within these limitations, choice will naturally fall on criteria believed to be most 'important', i.e. those which, if classification of the stands examined proves possible, are thought likely to lead to the most natural classification.

The principal criteria commonly used may be grouped under the following heads.

134

COMMUNITY DESCRIPTION AND COMPARISON

1. Floristic composition—the species present.
2. Measures of abundance of species.
3. Performance of individuals of species. This may be assessed either qualitatively, e.g. setting viable seed or not, or quantitatively, e.g. proportions of tillers flowering, mean height of plants, etc.
4. Growth and life form. These concepts are familiar enough, though difficult of precise definition. They represent attempts to classify individuals on the basis not of their taxonomic affinities, but of vegetative morphology.* Indeed, they may cut right across taxonomic affinity, as in the placing of succulent species of *Euphorbia* and members of the Cactaceae in the same class. Growth or life form grouping in relation to vegetation description may be comprehensive, assigning every species to its appropriate class in some such system as that of Raunkiaer, or it may be concerned with the occurrence of certain distinctive features of vegetative form, e.g. thorniness, formation of stilt roots, etc.
5. Physiognomy. The physiognomy of vegetation is likewise difficult of precise definition; it refers to the appearance of the stand as a whole and is closely connected with growth form, being largely influenced by the proportions of individuals of different growth forms; additional features depend on the performance and arrangement of individuals. It is the criterion least susceptible to exact description, though it is a most useful characteristic to an experienced ecologist. The terms for vegetation types used in common speech, such as woodland, moor, prairie, savannah, etc., mostly indicate physiognomy.
6. Pattern of the constituent species.
7. Various constants (in the mathematical sense) derived directly or indirectly from other criteria have been used, e.g. constants derived from the form of the species–area curve.

These criteria are not all independent of one another. Quantitative measures and measures of performance have meaning only between stands of broadly similar floristic composition. Growth form and physiognomy, on the other hand, are independent of floristic composition. Two stands of quite different floristic composition may have closely similar physiognomy and growth form spectra. Conversely, it is possible, at least theoretically, for two stands of the same floristic composition to differ in physiognomy and growth form spectrum, as some species differ in their growth form

---

* *Life form* and *growth form* have slightly different connotations. Life form is generally used where the classification is believed to have an adaptational significance, as in Raunkiaer's system, and growth form when no such significance is intended but different authors vary in their use of the terms (cf. Du Rietz, 1931).

135

according to the conditions in which they are growing. Pattern, although determinable only in terms of particular species, or groups of species, has meaning and can be compared independently of the species concerned; it is possible for two stands to show similar sets of patterns of the species composing them, although their floristic composition is quite dissimilar. The fundamental criterion in characterizing communities is the essentially subjective one of physiognomy. Only between stands of similar physiognomy are the more detailed and critical criteria relevant. In complex communities of several layers it may even be impossible to place the characterization of different layers on a common basis, e.g. the tree and ground layers of forest. The fundamental importance of physiognomy accounts for the breakdown of attempts to characterize and classify all stands according to a standard technique.

The handling of data on the different criteria may be considered for each in turn.

Floristic composition

The comparison of stands in terms only of the species present, without any reference to the abundance of species in the several stands, is a crude and insensitive mode of characterization. Only subjective comparison of species lists is, therefore, usually attempted, to determine whether or not a more laborious quantitative examination is necessary to assess the degree of similarity of two or more stands. Comparisons of species lists are, however, valuable on a much larger scale, the province more of the plant geographer than of the ecologist, e.g. in comparing the floras of different islands. In view of this, and because the principles underlying any attempt at an objective assessment of the degree of similarity between two species lists have not been fully realized in the past, it is worth while to consider the question a little further.

Jaccard, in a series of papers from 1902 onwards (e.g. see Jaccard, 1912), used a very simple *coefficient of community*, defined as the number of species common to the two areas expressed as a percentage of the total number of species, to assess the degree of floristic similarity between two areas. Jaccard's coefficient has been generalized by Koch (1957) as an *index of biotal dispersity*, to estimate the overall similarity of a number of species lists. If n lists include $s_1, s_2, \ldots s_n$, species respectively and the total number of species recorded is S, Koch's index is $\dfrac{100(T-S)}{(n-1)S}$, where $T = s_1 + s_2 + \ldots s_n$. If $n = 2$ this reduces to Jaccard's coefficient.

Another widely used coefficient is generally attributed to Sørensen

(1948) but, according to Curtis (1959), was originally proposed by Czekanowski (1913). This takes the form of the number of species common to the two areas expressed as a percentage of the mean number of species per area, i.e. $\dfrac{2c}{a+b} \times 100$ where the two areas contain a and b species respectively and there are c species in common.

In this notation Jaccard's coefficient is $\dfrac{c}{a+b-c} \times 100$. Various other indices of similarity between two species lists have been proposed (see Dagnelie, 1960) but have not been so widely used.

Jaccard assumed that if two areas were samples of the same vegetation, the coefficient would be 100 per cent. This approach, not unreasonable when it was put forward, appears naïve in the light of our present knowledge of the properties of small samples from a large population. If two samples of different size are taken from a uniform area of vegetation, we may expect to appear in the larger sample one or more rare species that are absent in the smaller; at least, this will apply unless both samples are very large. The sample area at which no further increase in number of species occurs, if such a size exists, will be considered below. Thus if two stands being compared are of different area a coefficient of community of 100 per cent is not to be expected even from stands representing the same community. This objection could be overcome by comparing species lists from similar areas within the two stands, but there is a further difficulty; even two samples of the same size, taken from 'identical' stands, will not, unless large enough to contain all the species of the hypothetical community of which the stands are themselves samples, include the same species. Thus, for species lists representing samples from the same population, the expected value of the coefficient is not 100 per cent but some lower figure.

The value of this figure will depend on the relationship between the number of species and the number of individuals in the vegetation under consideration. Williams has suggested that Fisher's logarithmic series applies to the relationship and has both tested observed data against expectations from the logarithmic series, and used the series in comparisons of species lists (see especially Williams, 1947, 1949). If n is the number of species represented by one individual only in a sample, then in the logarithmic series

$$n, \frac{n}{2}x, \frac{n}{3}x^2, \frac{n}{4}x^3, \text{ - - - },$$

the second term gives the number of species with two individuals, the third the number with three individuals, etc. x is a constant depending on the sample size. Fisher has shown that the ratio $\dfrac{n}{x}$ is constant for the same population, whatever the sample size. Williams has designated this constant α, the *index of diversity*. A high value indicates great diversity, i.e. relatively fewer individuals per species, a low value little diversity, i.e. relatively more individuals per species. The series may also be written

$$\alpha x,\ \alpha\ \frac{x^2}{2},\ \alpha\ \frac{x^3}{3},\ \alpha\ \frac{x^4}{4},\ \text{-- -- -- .}$$

It follows from the properties of the logarithmic series that if N is the number of individuals in any sample and S the number of species,

$$S = \alpha \log_e \left(1 + \frac{N}{\alpha}\right).$$

If N is large compared with α, as it normally is in comparison of vegetation stands, the 1 may be neglected and this expression reduces to

$$S = \alpha \log_e \left(\frac{N}{\alpha}\right).$$

In vegetation the individual is an uncertain unit, as we have seen in discussing density. Williams suggests that area may be substituted for number of individuals, it being reasonable to suppose, for similar types of vegetation, that the total number of 'plant units' per unit area is constant. Consider two samples, of areas A and B bearing a and b species respectively. Let the total number of species represented, on the assumption that the two samples are drawn from a single population, be T and the mean number of individuals per unit area be I. Then from the relationship above

$$T = \alpha \log_e \left[\frac{I(A + B)}{\alpha}\right]$$

$$a = \alpha \log_e \left(\frac{IA}{\alpha}\right)$$

$$T - a = \alpha \log_e \left[\frac{I(A + B)}{\alpha}\right] - \alpha \log_e \left(\frac{IA}{\alpha}\right)$$

$$= \alpha \log_e \left(\frac{A + B}{A}\right).$$

Similarly $T - b = \alpha \log_e \left(\dfrac{A + B}{B}\right).$

From these equations, knowing A, B, a, and b, it is possible to calculate T and α.

If the samples are drawn from a single population, the number of species in common will be $a + b - T$, and the observed number may be compared with this. Williams quotes the example of the islands of Guernsey, 24 square miles in area with 804 species, and Alderney, 3 square miles in area with 519 species, which have 480 species in common. The equations are

$$T - 804 = \alpha \log_e \left(\frac{27}{24}\right)$$

$$T - 519 = \alpha \log_e \left(\frac{27}{3}\right)$$

from which $T = 820$. The expected number of species in common is, therefore, $804 + 519 - 820 = 503$. The observed number, 480, is 95·4 per cent of this, indicating a close similarity between the two floras. Jaccard's coefficient is

$$\frac{480 \times 100}{804 + 519 - 480} = \frac{480 \times 100}{943} = 50·9 \text{ per cent.}$$

Preston (1948) has put forward the alternative hypothesis of a lognormal relationship between the number of species and of individuals, with the number of species in successive 'octaves' (groups delimited by successive doubling of the number of individuals, i.e. <1, 1–2, 2–4, 4–8, etc.) forming a normal distribution. The lowest groups may be lacking in a small sample, being cut off by what he terms a 'veil line', which shifts to the left with increasing sample size. Thus, if a sample is small relative to the population it represents, there will be a greater number of species represented by a single individual than by two individuals, as with the logarithmic series, but if the sample is large relative to the population this will not necessarily apply. Preston himself tested only animal populations. Although Black et al. (1950) presented data from plots in Amazonian forest in the form of 'octaves' of density, no test has apparently been made of plant populations. The expected number of species common to two lists, on the assumption that they represent samples from the same lognormal population, has been derived by Preston (1962). If N_a and N_b are the numbers of species in two samples of the population and N_{a+b} is the number of species in the combined area, then

$$(N_{a+b})^{\frac{1}{z}} = (N_a)^{\frac{1}{z}} + (N_b)^{\frac{1}{z}}$$

where z is a constant derived from the lognormal distribution. Preston shows that if the two samples are drawn from the same lognormal universe there are theoretical grounds for believing that z is approximately 0.27. Applying the relationship to the data quoted the expected total number of species

$$T = \left\{(804)^{\frac{1}{0.27}} + (519)^{\frac{1}{0.27}}\right\}^{0.27} = 844,$$

giving as the expected number of species in common $804 + 519 - 844 = 479$, a value nearly identical with that observed.*

Both hypotheses are dependent on various unproved assumptions, and their validity can only be tested by their degree of correspondence with observed values. Unfortunately, as Goodall (1952a) remarks, the differences between the two sets of expectations, for most sets of data, are not sufficient to permit a decision. Both hypotheses assume a random distribution of individuals, which, while perhaps not unreasonable for geographical comparisons such as the example quoted, will certainly not be satisfied in comparisons of stands. It may be noted in relation to geographical comparisons that Preston's lognormal hypothesis, which postulates moderately common species as more numerous than either very rare or very common ones, perhaps accords better with the general experience of field botanists than does Williams's logarithmic hypothesis, which assumes very rare species to be the most numerous.

It is clear that any attempt at objective comparison of species lists from stands (however useful it may be for comparison of floras of different regions) is based on so many unproved assumptions that it is scarcely worth attempting. Especially is this so in view of the crudity and imprecision of species lists as criteria in characterizing stands.

Quantitative measures of abundance and performance

The appropriate methods of sampling in determination of measures of abundance and performance have already been considered in Chapter 2. Comparison of mean values of a measure of a single species is, as we have seen, quite straightforward. We are here concerned with assessing the total difference between two stands when all the constituent species are considered together. This is a more complex problem, the solution of which has been indicated by Hughes and Lindley (1955), who point out the advantages of the statistic D^2 of Mahalanobis and Rao (p. 123).

* Reference may be made to Aitchison and Brown (1957) for an extended account of the mathematical aspects of the lognormal distribution.

They have applied it to vegetation types and quote data on oak scrub, dominated by hybrids of *Quercus robur* and *Q. petraea*, and found both on acidic soils derived from rhyolite and basic soils derived from pumice tuff. The accompanying species were assessed in a number of stands according to 'prominence' on a scale of 1–10. The species were grouped into four classes according to their ecological preferences and the mean prominence for each group determined for stands on acidic and basic soils respectively (TABLE 12). D^2 between the fifteen stands on acidic soil and the 22 on basic soil is 11·647785.* The probability of getting by chance this or a

TABLE 12

DIFFERENCES BETWEEN THE COMPOSITION OF TWO TYPES OF OAK SCRUB IN THE CONWAY VALLEY, NORTH WALES (from Hughes and Lindley, 1955, by courtesy of *Nature, Lond.*)

Ecological class of species	Mean prominence (scale 1–10) of species class		Difference $(R–P)$
	R Soils derived from rhyolite (acidic) (15 samples)	P Soils derived from pumice tuff (basic) (22 samples)	
A. Species of *Quercetum roboris*	1·28	4·09	−2·81
B. Species of *Quercetum petraeae*	2·80	2·28	+0·52
C. Species of heaths and acidic grasslands	2·90	0·44	+2·46
D. Species of agricultural grasslands	2·42	3·07	−1·65

greater value in two sets of samples drawn from a population homogeneous in terms of prominence of the four species groups is less than 1%. The conclusion is drawn that the vegetation developed on the two soil types shows a real difference in composition.

The approach, which has, so far, been used only in this one case, is worth further consideration. In this example a number of different stands, falling naturally into two groups on the basis of soil type, was examined and a single abundance figure attached to each species for each stand. Measures from a number of samples within each of two stands could be treated in the same way, to give a D^2 comparing the two stands. However, the computational labour of calculation of D^2

* Hughes and Lindley (personal communication) now prefer this value to be quoted as 11·65.

is such that it may rarely seem worth while to use it for two single stands. If there is a *prima facie* case, as with the oak scrub, for grouping the stands into classes, the means, variances and covariances for the total data from a group of stands may be grouped and used as were those for the single measures per stand analysed by Hughes and Lindley. Density, yield, etc., could readily be handled for two stands alone, but frequency and cover could not, as, though there is available an estimate of the variance of frequency (or cover) from random quadrats or points within a stand, there is no measure of covariance between one species and another.

Hughes and Lindley grouped their species into four classes of different ecological preferences. This must necessarily have been done subjectively and the criticism might be made that an initial bias was introduced in this way. There is no theoretical reason why each species should not be treated separately but in practice the amount of computation for more than a few species would be prohibitive. The number of variables considered must, therefore, be kept small. In the case quoted the grouping of species by ecological preference was probably simple, granted an initial field knowledge of similar vegetation. This will not always be so and an alternative procedure may be considered of analysing only the five or six most abundant species. Although this deliberately ignores evidence derived from some, and possibly many, species, it may be preferable to a dubious subjective grouping of the species into ecological classes. If the species are grouped, a measure that does not distort the relative prominence of different species must be used. Thus, in many types of vegetation, density would not be satisfactory.

Various workers have attempted, by combining two or more measures into a single item, to make a more comprehensive estimate of the importance of species in a stand than is given by any one of the measures of abundance. For example, Curtis (1947, and see Brown and Curtis, 1952; Curtis and McIntosh, 1951) used an *importance value* obtained by adding together relative frequency (frequency of species as percentage of total frequency values of all species), relative density, and relative 'dominance' (basal area for species as percentage of total basal area). Such derived measures may be handled in comparisons in the same way as simple measures. They are open to the criticism that the mode of combining is quite arbitrary and markedly different situations may give rise to the same combined value. In the present case, if density is constant, the same importance value may be given by a regular distribution of saplings (high frequency, low basal area) as by groups of large trees (low frequency, high basal area). However, the use of single measures, especially density and frequency, may likewise fail to detect gross differences in the make-up of the vegetation. A combined measure should not be used uncritically; its relative advantages and disadvantages should be considered in relation to the vegetation being examined.

An alternative approach to comparison of communities in terms of abundance of different species is to use an index of similarity of species composition in which the contribution of each species observed is weighted by a measure of its abundance. While less precise than the techniques discussed above, this involves much less computation and permits the use of much less extensive data on the communities being compared, e.g. a single set of frequency estimates for each community. The weighting of species by a measure of abundance was suggested by Gleason (1920) and has been sporadically used since. More recently, however, such weighted indices have been more critically considered in view of their importance in many of the techniques of classification and ordination to be discussed in the next chapter.

Both the Jaccard and Sørensen coefficients of community may be weighted in this way, as may most other coefficients (see Dagnelie, 1960). A useful modification, applied by various workers, is to express species abundance not as the absolute values measured, but relatively to the total abundance of all species in a community, i.e. as percentage contribution to the vegetation. This is, in most circumstances, ecologically more realistic as it eliminates the effect of differences in the overall performance of the vegetation. It should not, however, be accepted uncritically for all cases as, at least theoretically, circumstances may occur in which relative contribution of different species is the same but variation in absolute amount of vegetation is ecologically significant; this might apply, for example, in studies of populations of ephemerals in relation to available moisture. When percentage contribution is used to weight the species, Sørensen's coefficient reduces to the simple form of the sum of the lesser of the two values for each species, as Bray & Curtis (1957) have pointed out.

Morisita (1959b) has suggested a coefficient which is also independent of absolute density. This takes the form

$$\frac{2\Sigma n_{1i}n_{2i}}{(\lambda_1 + \lambda_2)N_1 N_2}$$

where n_{1i}, n_{2i} are the number of individuals of the ith species, N_1, N_2 the total number of individuals in the two communities, and λ_1, λ_2 Simpson's measure of diversity (see p. 67) for occurrence of individuals of different species in the two communities. Morisita suggests similar coefficients for other absolute measures. This coefficient involves considerably heavier computation than the Sørensen coefficient based on the relative measures and, at least where large numbers of coefficients are required, does not appear to have compensatory advantages.

The apparently straightforward use of a correlation coefficient between the amounts of different species in two communities as a measure of similarity (Motomura, 1952) has serious disadvantages. The coefficient has a low sensitivity at relatively high values, where the ecological significance is generally greatest (Bray & Curtis, 1959). Further, data of this type are not generally normally distributed, and the use of the correlation coefficient as a measure of similarity is dependent on normal distribution of the data (cf. p. 108). Ghent (1963) has suggested using a rank correlation coefficient (see p. 103), which is free from these objections. This suggestion is worthy of a more extensive trial.

Growth and life form

Data on growth or life form usually embrace a record of the number of species falling into each class of some suitable classification, e.g. that of Raunkiaer, or, if certain morphological features only are the main interest, the number of species with and without each feature. These data are based on the species list for the stand, and, like it, provide a relatively insensitive criterion. It is of greater value when applied to regions rather than to stands. Although data are commonly presented as a *spectrum*, giving the percentage of species in each class, the actual number of species of each class in different sets of data must be used in objective comparisons, which can readily be made by analysis of contingency tables.

TABLE 13 shows data, from Ewer (1932), of the number of species of different life forms according to Raunkiaer's classification, in four regions of the State of Illinois. (A very small number of epiphytes and stem succulents is omitted.) The percentages of species having the different life forms are similar, and the question is posed whether or not there is any real difference between the life form spectra of the four regions. If there is no difference, the numbers in different classes in each region should be in the same proportion as the totals shown in the bottom row of the table. The expected values and the χ^2 contributions resulting from the difference between observed and expected are shown in each cell of the table. The total χ^2 of 51·7596 has 24 degrees of freedom (8 $\times$ 3 from a

TABLE 13

NUMBERS OF SPECIES OF DIFFERENT LIFE FORMS IN FOUR REGIONS OF ILLINOIS (Data of Ewer, 1932)

Life form

Region		Megaphanerophyte Mg	Mesophanerophyte Ms	Microphanerophyte Mc	Nanophanerophyte N	Chamaephyte Ch	Hemicryptophyte H	Geophyte G	Helophyte and Hydrophyte HH	Therophyte T	Total
North	Obs.	18	51	82	58	19	686	192	68	193	1367
	Exp.	24·00	70·13	75·11	40·52	13·71	692·85	180·15	62·33	208·20	
	χ^2	1·5000	5·2183	0·6320	7·5407	2·0411	0·0678	0·7795	0·5158	1·1097	
Central	Obs.	16	52	58	30	10	637	165	60	198	1226
	Exp.	21·52	62·89	67·37	36·34	12·30	621·39	161·57	55·91	186·72	
	χ^2	1·4159	1·8857	1·3032	1·1061	0·4301	0·3921	0·0728	0·2992	0·6814	
Mid-South	Obs.	20	55	45	24	6	454	112	31	142	889
	Exp.	15·61	45·61	48·85	26·35	8·92	450·58	117·16	40·54	135·40	
	χ^2	1·2346	1·9332	0·3034	0·2096	0·9559	0·0260	0·2273	2·2450	0·3217	
South	Obs.	23	67	56	18	9	446	109	41	135	904
	Exp.	15·87	46·37	49·67	26·79	9·07	458·18	119·13	41·22	137·68	
	χ^2	3·2033	9·1783	0·8067	2·8841	0·0005	0·3238	0·8614	0·0012	0·0522	
Total		77	225	241	130	44	2223	578	200	668	4386

9 × 4 table) and a probability of slightly less than 0·1 per cent. There is evident discrepancy among the four areas. Inspection of the table suggests that there is a greater proportion of phanerophytes in the two southern areas and also that the proportions of the different types of phanerophytes vary from area to area. The data may be divided to test these points. TABLE 14 shows the number of

TABLE 14

As TABLE 13. NUMBER OF SPECIES IN PHANERO-
PHYTE AND NON-PHANEROPHYTE CLASSES

Life form

Region		Phanerophyte	Non-phanerophyte	Total
North	Obs.	209	1158	1367
	Exp.	209·76	1157·24	
	χ^2	0·0028	0·0005	
Central	Obs.	156	1070	1226
	Exp.	188·12	1037·88	
	χ^2	5·4842	0·9940	
Mid-South	Obs.	144	745	889
	Exp.	136·41	752·59	
	χ^2	0·4223	0·0765	
South	Obs.	164	740	904
	Exp.	138·71	765·29	
	χ^2	4·6109	0·8357	
Total		673	3713	4386

phanerophytes and 'non-phanerophytes' in the four areas. It gives a total χ^2 of 12·4269 with three degrees of freedom and probability between 0·1 and 1 per cent. TABLE 15 is concerned with the phanerophyte classes alone. (Note that the expected values are different, owing to the different proportions of total phanerophytes in the four regions.) χ^2 is 28·0006 with 9 degrees of freedom and probability slightly less than 0·1 per cent. Examination of the differences between expected and observed values shows a regular

146

increase in the proportion of the megaphanerophyte and mesophan-
erophyte classes towards the south. By contrast, the table for the non-
phanerophyte classes (TABLE 16) has a total χ^2 of 9·5128 with
12 degrees of freedom and probability 50–70 per cent. There is
no indication of difference in proportion of life forms other than
phanerophyte. Here, as generally with data involving assignment
of units to one of a number of classes, χ^2 contingency analysis
provides a flexible technique readily adapted to particular cases.

TABLE 15

As TABLE 13. PHANEROPHYTE CLASSES ONLY

Life form

Region		Mg	Ms	Mc	N	Total
North	Obs.	18	51	82	58	209
	Exp.	23·91	69·87	74·84	40·37	
	χ^2	1·4608	5·0963	0·6850	7·6992	
Central	Obs.	16	52	58	30	156
	Exp.	17·85	52·15	55·86	30·13	
	χ^2	0·1917	0·0004	0·0820	0·0006	
Mid-South	Obs.	20	55	45	24	144
	Exp.	16·48	48·14	51·57	27·82	
	χ^2	0·7518	0·9776	0·8370	0·5245	
South	Obs.	23	67	56	18	164
	Exp.	18·76	54·83	58·73	31·68	
	χ^2	0·9583	2·7012	0·1269	5·9073	
Total		77	225	241	130	673

In small samples, e.g. particular stands, it is more informative to
consider the abundance of plants of different growth (or life) forms
rather than the number of species. This may be done by calculating
a growth form spectrum in which the contribution of each species
to its growth form class is weighted by some measure of abundance.
This gives data which are not readily compared and any objective
comparison must be of the actual numbers of individuals of different
growth forms, or numbers of points or quadrats touched or occupied
by plants of different growth forms.

Physiognomy

As already seen physiognomy is the criterion least susceptible
to any quantitative approach and must usually rest on verbal

description (on descriptive symbols see Christian and Perry, 1953). The only objective approach that has proved of value is the profile diagram introduced by Richards for forest communities (Davis and Richards, 1933–4; see also Richards, 1952). This is prepared by careful measurement of height, girth, crown, etc., of all the individuals in a strip of forest. These individuals are then drawn conventionally, but to scale for the measured features, in their

TABLE 16

As Table 13. Non-phanerophyte Classes Only

Life form

Region		Ch	H	G	H H	T	Total
North	Obs.	19	686	192	68	193	1158
	Exp.	13·72	693·30	180·27	62·38	208·33	
	χ^2	2·0320	0·0769	0·7633	0·5063	1·1281	
Central	Obs.	10	637	165	60	198	1070
	Exp.	12·67	640·62	166·57	57·64	192·50	
	χ^2	0·5627	0·0205	0·0148	0·0966	0·1571	
Mid-South	Obs.	6	454	112	31	142	745
	Exp.	8·83	446·04	115·97	40·13	134·03	
	χ^2	0·9070	0·1421	0·1359	2·0772	0·4739	
South	Obs.	9	446	109	41	135	740
	Exp.	8·77	443·04	115·20	39·86	133·13	
	χ^2	0·0060	0·0198	0·3337	0·0326	0·0263	
Total		44	2223	578	200	668	3713

correct positions along a line representing the length of the strip. Such diagrams give a clear impression of many of the features covered by physiognomy.

Pattern

Although detection and description of pattern, unless it is very pronounced, is dependent on a quantitative approach, there is no entirely satisfactory method of describing pattern within a stand by any simple expression, even for one species. Various measures of departure from randomness based on data from a single size of quadrat have been proposed, but are all more or less unsatisfactory. A more satisfactory description of pattern is obtained from a curve of variance against sample size. (Chapter 3.) Such curves refer only to individual species and cannot be integrated for a stand.

Comparison of 'total pattern' of stands must, therefore, be confined to a subjective assessment of the sets of curves obtained from each.

Community constants

Four community characteristics derived from quantitative data have been the subject of considerable discussion. These are (1) the frequency distribution of species, (2) the generic coefficient, (3) the form of the species–area curve and, perhaps the most discussed, (4) the minimal area of a community. *Frequency distribution*, the proportion of species in the stand having various frequency values, has already been discussed in Chapter 1, where it has been shown that its form is dependent on a complex of factors, including species diversity in relation to number of individuals, pattern and, especially, quadrat size. It is correspondingly difficult to interpret and therefore of little general value.

The *generic coefficient* of a sample, i.e. the number of genera expressed as a percentage of the number of species, was introduced by Jaccard (see Jaccard, 1912), and its value has been regarded alternatively as providing information on the variety of habitats, or on the relative intensity of intergeneric and intrageneric competition in the area. Its worth is limited, however, by its dependence not only on the generic diversity of the population from which the sample is drawn but also on the size of sample (in terms of number of species) used. Without assuming any particular relationship between number of genera with 1, 2, 3, etc. species represented it is readily shown that as the sample size increases the generic co-efficient falls. Suppose a population consists of a, b, c, d, etc., genera represented by 1, 2, 3, 4, etc., species respectively. Then for the whole population the generic coefficient is

$$G = \frac{a + b + c + d + \ldots}{a + 2b + 3c + 4d + \ldots}.$$

If a sample contains a proportion x of the total number of species, it may be expected that the same proportion x will be drawn from species belonging to genera having 1, 2, 3, etc., species in the population. Thus a number ax of genera with one species in the population will be represented. Some of the genera containing two species will be represented by both species, some by one, some by none, the relative proportions being given by the binomial expansion $[x + (1 - x)]^2$, so that $b(1 - x)^2$ genera will not be represented, and $b[1 - (1 - x)^2]$ will be represented. Similarly $c[1 - (1 - x)^3]$, $d[1 - (1 - x)^4]$ etc. of the genera containing 3, 4, etc., species will be represented. Thus the number of species in the sample is $x(a + 2b + 3c + 4d + \ldots)$, the number of genera $ax +$

149

$$b[1 - (1 - x)^2] + c[1 - (1 - x)^3] + d[1 - (1 - x)^4] + \cdots,$$

and the generic coefficient

$$C =$$
$$\frac{ax + b[1 - (1 - x)^2] + c[1 - (1 - x)^3] + d[1 - (1 - x)^4] + \cdots}{x(a + 2b + 3c + 4d + \cdots)}$$

Let $S = a + 2b + 3c + 4d + \cdots$

Then
$$C = \frac{1}{S}\left\{ \frac{a[1 - (1 - x)]}{x} + \frac{b[1 - (1 - x)^2]}{x} \right.$$
$$\left. + \frac{c[1 - (1 - x)^3]}{x} + \frac{d[1 - (1 - x)^4]}{x} + \cdots \right\}$$
$$= \frac{1}{Sx}(a + b + c + d + \cdots) - \frac{1}{S}\left[\frac{a(1 - x)}{x} + \frac{b(1 - x)^2}{x} \right.$$
$$\left. + \frac{c(1 - x)^3}{x} + \frac{d(1 - x)^4}{x} + \cdots \right]$$
$$= \frac{G}{x} - \frac{1}{S}\left[\frac{a(1 - x)}{x} + \frac{b(1 - x)^2}{x} + \frac{c(1 - x)^3}{x} \right.$$
$$\left. + \frac{d(1 - x)^4}{x} + \cdots \right]$$

Differentiating
$$\frac{dC}{dx} = -\frac{G}{x^2} - \frac{1}{S}\left[-\frac{a(1 - x)}{x^2} - \frac{a}{x} - \frac{b(1 - x)^2}{x^2} \right.$$
$$- \frac{2b(1 - x)}{x} - \frac{c(1 - x)^3}{x^2} - \frac{3c(1 - x)^2}{x} - \frac{d(1 - x)^4}{x^2}$$
$$\left. - \frac{4d(1 - x)^3}{x} - \cdots \right]$$
$$= -\frac{G}{x^2} + \frac{1}{S}\left[\frac{a}{x^2} + \frac{b(1 - x)(1 + x)}{x^2} + \frac{c(1 - x)^2(1 + 2x)}{x^2} \right.$$
$$\left. + \frac{d(1 - x)^3(1 + 3x)}{x^2} + \cdots \right]$$

If $x < 1$, the expressions $(1 - x)(1 + x)$, $(1 - x)^2(1 + 2x)$, etc. are also less than one.

Hence
$$\frac{dC}{dx} < -\frac{G}{x^2} + \frac{1}{S}\left(\frac{a}{x^2} + \frac{b}{x^2} + \frac{c}{x^2} + \frac{d}{x^2} + \cdots \right)$$
$$= -\frac{G}{x^2} + \frac{G}{x^2} = 0$$

i.e. within the range $0 < x < 1$, $\frac{dC}{dx}$ is negative and C decreases with increasing x.

Maillefer (1929) established empirically this decrease in the generic coefficient with increasing sample size, by means of random samples from the Swiss flora. Thus, no simple interpretation of differences in the values for two communities can be made. If a definite hypothesis of the relationship between number of species and number of genera, such as the logarithmic relationship proposed by Williams, is adopted, then the effect of sample size can be taken into account. In this case, however, the constant α of the logarithmic series gives a measure of generic diversity alone and is more satisfactory than the generic coefficient (Williams, 1949).

If the number of species present is recorded from a series of samples of increasing size in a stand and plotted against the areas sampled, a curve is obtained, the slope of which is at first steep but which gradually decreases. The form of such *species–area curves* has provoked much interest and various attempts have been made to fit the observed curves to equations, in the hope that the constants of the equations will represent characteristic features of the stands investigated. If this were so it might be possible to use the values of the constants to group stands into classes characterized by similarities of structure, regardless of difference in specific composition. It is not necessary here to review the extensive literature on species–area curves, which has been well summarized by Goodall (1952a) (see also Hopkins, 1955), but rather to attempt to assess their practical value in comparing communities.

Within one stand, area can be regarded as a measure of the number of plant units so that the form of the species–area curve is a reflection of the relative numbers of individuals and of the pattern of the different species present. The influence of pattern on the species–area curve has commonly been overlooked or ignored, although it is clear that if most species are markedly contagiously distributed, the number of species observed in a sample of a particular size will be less than if the species were randomly distributed, unless the sample size exceeds the maximum scale of heterogeneity of all the species present. Hopkins (1955) has demonstrated this effect. (See also Blackman, 1935.) Thus, some discrepancy from any theoretical curve based on random distribution of individuals is to be expected for this reason alone. A further source of discrepancy, as Goodall (1952a) has emphasized, may arise from the method of collecting data used. Three methods have been employed. (1) Separate samples, each randomly placed, are examined for each area to be used in the curve. (2) Each sample size is obtained by adding a contiguous area to the previous one. (3) The various sizes of sample are made up by adding the data from the appropriate number of randomly placed small quadrats. Only the first is satisfactory and, unfortunately, has been used least. In the second the data for different sizes are not independent. If a rare species appears in a sample of one size, it is necessarily present in all larger samples, so that the number of species tends to be exaggerated, especially at the smaller sample sizes. The third method shares the disadvantage of lack of independence between sizes. It further tends, as Blackman (1935) has shown, to eliminate the effect of contagious distribution, as two aggregated species, confined to different parts of the stand so that they could only occur together in a very large quadrat, may be included in quite a small size of sample.

Two hypotheses of the relation between number of species and number of individuals have been put forward, the logarithmic series and the lognormal distribution (pp. 137-40). If the corresponding species–area curves are plotted on a semilog basis (number of species against logarithm of area), they differ in shape. The relationship given by the logarithmic series is $S = \alpha \log_e \left(1 + \dfrac{IA}{\alpha} \right)$, where S is the number of species on an area A, I is the number of individuals per unit area, and α the index of diversity. This gives a curve at first having slope increasing with area and becoming effectively linear when 1 becomes negligible compared with $\dfrac{IA}{\alpha}$, so that the relationship simplifies to $S = \alpha \log_e \dfrac{IA}{\alpha}$, or $S = \alpha \log_e \left(\dfrac{I}{\alpha} \right) + \alpha \log A$.

Various authors, from Gleason (1922) onwards, have suggested, or accepted, on empirical grounds a relationship of the type $y = a + b \log x$ where y is number of species on area x and a and b are constants. If this hypothesis is accepted, the index of diversity thus obtainable from species–area data provides a constant of some biological interest, as expressing the richness of the flora of the area. Hopkins (1955) further points out that, if the linear part of the curve is projected backwards, it cuts the area axis at the point $\dfrac{\alpha}{I}$ ($\dfrac{\alpha}{N}$ in his terminology), which 'may be regarded as an objectively defined area of a plant community'. This area is small, ranging from 0·6 to 24,000 sq. cm in the communities he studied, and he does not suggest any biological significance for it.

As Preston has shown, the lognormal distribution leads to a sigmoid species–log-area curve. Vestal (1949) claimed that data from a large number of communities have sigmoid species–log-area curves, but the species–area data for nearly all the communities were 'reconstructed' from published accounts by a dubious method. Archibald (1949b) has also proposed a sigmoid relationship, approximated by

$$\log_{10} \frac{y}{S-y} = k \left(\log_{10} x - \log_{10} x_{50} \right)$$

where y is the number of species on area x, x_{50} is the area on which an average of half the total number of species in the community occurs, S is the total number of species, and k is a constant. This relationship, though supposedly based on the assumption of random distribution, is, in fact, as Goodall (1952a) has pointed out, based on regular distribution, a state of affairs still further from that found in the field than the random distribution assumed by Williams and Preston. Moreover, none of Archibald's data, with one doubtful exception (*Limonium* marsh, examined further by Hopkins, 1955), includes any part of the upper, convex, part of the supposed sigmoid curve.

There has been considerable but rather sterile debate as to whether the upper parts of species–area curves are in fact linear or sigmoid, though the balance of opinion is in favour of the linear form. The difficulty of decision arises from the fact that the suggested sigmoid curves are very nearly linear up to the largest areas normally included in species–area observations. In view of the great uncertainty about the form of the curves, constants derived from equations supposedly fitting them must be regarded with grave suspicion. The approximately linear form of the species–log–area curve over the ranges of area normally met with does, however, mean that the slope of the line can usefully be used as an empirical measure of the relative species diversity of communities.*

Closely related to the form of the species–area curve are various concepts of *minimal area* and *representative area*. Though varying terms and definitions have been applied to these concepts, they have a common basis in the idea that the true characteristics of a plant community only appear when a certain minimum area of it is examined. In broad terms this is a truism; it is evident that the larger the sample examined the greater will be the information

*The only critical examination of species–area relations for large areas appears to be Preston's (1960) consideration of the breeding birds of the nearctic and neo-tropical regions. He concluded that, although the total species–log–area curve was sigmoid, there was a close approximation to a straight line over the range 10–100,000 or even 10–1,000,000 acres. In the absence of the points for still larger areas "we should undoubtedly 'graduate' it as a straight line and believe we had proved the point". There is no reason to suppose that Preston's argument necessarily applies to species–area data for plants, but it does bring out the danger of basing conclusions on a limited range of areas.

obtained, whether about the species present, their quantitative representation or their patterns. Likewise a law of diminishing returns will apply in that successive equal increases in size of sample will give successively smaller amounts of additional information. The validity of determining, other than empirically, any precise area to be used in describing vegetation depends on whether or not there is, at some point, a sudden decrease in the amount of additional information obtained. Goodall (1952a) remarks that such concepts as minimal area have little significance outside a system of classification. This appears to be a rather unfortunate side-tracking of the issue. If concepts of characteristic areas do have any validity, they must have an important bearing on any comparison of stands, whether the latter can be placed in a system of classification or not.

As in consideration of species–area curves, it is not necessary to review the rather extensive literature; reference may again be made to Goodall's (1952a) comprehensive review. There have been three approaches to the determination of characteristic areas, based respectively on species composition, species frequency, and homogeneity of composition. The species–area curve, as we have seen, at first rises sharply and then flattens off, eventually becoming nearly horizontal. The area above which it becomes nearly horizontal has been regarded by various ecologists as an indication of the minimal area. (See Braun-Blanquet, 1951, for a recent exposition of this point of view.) Cain (1934) claimed that there is a 'break' in the curve at which its slope decreases sharply, and many curves from field data do show this 'break'. Cain (1938) himself later realized that the position of the break depends on the relative scales of the ordinate and abscissa. Realization of this difficulty led to attempts to determine a characteristic area from the species–area curve in some other way. Cain (1938) suggested the point where the slope was equal to the average slope for the whole curve, or to some fraction of it (Cain, 1943). The value obtained, however, depends on the largest area examined, as does Archibald's (1949b) use of the area containing 50 per cent of the total number of species. This objection does not apply to Vestal's (1949) use of a pair of points, such that the ratio of the corresponding areas is 1 : 50 and of the species numbers 1 : 2, to define the minimal area as five times that corresponding to the lower of the pair of points. These points are readily determined, particularly if a graph is drawn on a log basis, but it is difficult to see their particular significance in relation to the vegetation. Even if this concept of minimal area is accepted, it is evident that its value for a particular stand can only be determined very approximately. What has been said above on the species–area curve and on attempts to determine a minimal area from it does not detract from the value in some circumstances, e.g. in species-rich tropical forest, of plotting such a curve as a guide as to whether most of the species present have been recorded or not.

Archibald (1949a) has suggested the minimum quadrat size which shows 95 per cent frequency for at least one species as a minimal area (with particular reference to comparison of 'dominants'). The principal elaboration of a minimal area concept based on frequency is that of the Uppsala school, e.g. Du Rietz (1921). It is claimed that, if frequency is estimated with increasing size of quadrat, and if the number of species having a frequency of more than 90 per cent is plotted against quadrat size, the curve becomes and remains horizontal, at least for a considerable further increase in area. Those species with frequency greater than 90 per cent are referred to as constants and the minimal area is defined as that at which the full number of constants is attained or, at least, the number of constants remains the same for a considerable further increase in area. The crux of this approach is the doubtful validity of the assumed relationship between the number of constants, as defined, and area. There has been marked disagreement over this (see Pearsall, 1924; Goodall, 1952a); it is relevant to point out that the data put forward in support of the supposed relationship were obtained from quadrats selectively placed and not from random sampling.

Both the scale and intensity of pattern of the different species present will affect the size of minimal area found, if, indeed, any definite area can be determined. A species will appear at a relatively small size of quadrat if the only pattern it exhibits is small scale. If large-scale pattern is present, the size at which it appears will depend on its intensity; a species with dense clumps separated by spaces in which it is absent (high intensity of pattern) will tend to appear only in large quadrats. Conversely, a species with large-scale pattern of a mosaic of patches with higher and lower density (low intensity of pattern) will appear at a smaller quadrat size. Minimal area is thus not related to pattern in any simple manner; the same minimal area may be found for stands of quite different structure. The concept has, therefore, little practical value.

Goodall (1954b) has put forward a different approach to minimal area, based on homogeneity of values of a quantitative measure of the various species in replicate samples at different spacings. He suggested as the minimal area 'the smallest sample area for which the expected differences in composition between replicates are independent of their distance apart', i.e. are due to random variation only. He presented data for salt marsh in support of the view that no such area can be found or, at least, that it is greater than the largest area examined (10 sq. metres). Later he found some evidence for a minimal area of a quarter of a mile square in Uganda rain forest but no evidence for minimal area in mallee nor in more extended samples from salt-marsh (Goodall, 1961). This approach is, in effect, a direct examination of scale of patterning and attempts to determine the maximum scale of pattern present for each species and hence that for all species in the stand. The data presented are in a form similar to that used by Greig-Smith (1952a, etc.), Kershaw (1957a, etc.) and others, and permit an estimate of the several scales of heterogeneity present (Greig-Smith et al., 1963). The minimal area, as defined, refers to a part only of the information obtained and it scarcely seems desirable to single it out for special consideration.

We have been considering minimal area as one characteristic of the community. Before leaving the subject it may be well to emphasize that, in spite of some confusion in the literature, it bears no necessary relationship, however it is determined, to the most suitable size of sample area for determination of other characteristics of the community. The most suitable quadrat size for determination of frequency, or density, or any other quantitative measure of species rests upon the considerations of edge effect, ease of recording, minimizing of variance of mean, and shape of distribution curve obtained, matters already dealt with in Chapter 2.

In concluding this chapter it is convenient to refer again to consideration of the homogeneity of communities. This may refer either to homogeneity within a stand, discussed above (p. 133), or to the degree of homogeneity shown by the various stands assigned to an abstract vegetational unit. When comparing the vegetational units in different systems of classification it may be useful to assess their homogeneity as an indication of how broadly the units are defined in the different systems. In such assessments of homogeneity attention has mainly been concentrated on specific composition. A simple measure of homogeneity is provided by the mean value, for all possible comparisons between stands, of one or other of the coefficients of community. Curtis (1959) has suggested as an index the ratio of total presence value of the prevalent species to total presence value of all species. (Prevalent species are defined as those species, to the number of the average species density per stand, with the highest presence value.) Dahl (1957, 1960) has pointed out that, if the logarithmic distribution of individuals between species is accepted, Curtis' index and the Jaccard and Sørensen coefficients of community can be determined in terms of α, the index of diversity, and the mean number of species per stand. He suggests as an index of uniformity the ratio of mean number of species per sample to the index of diversity, higher values indicating greater homogeneity. If it is desired to estimate homogeneity in terms of the quantities of different species, coefficients of community based on weighted values may be used (p. 143).

CHAPTER 7

PLANT COMMUNITIES—II. CLASSIFICATION AND ORDINATION

HAVING considered in the last chapter the techniques of describing and comparing stands we may turn to techniques of summarizing and ordering the available information. To do so is important not only as an end in itself but as a basis of comparison with the environmental and other factors affecting the composition of vegetation. This in turn leads to an assessment of the causation of the composition of stands as we find them.

In recent years a cleavage has appeared between two different approaches to this problem. *Classification* involves arranging stands into classes, the members of each of which have in common a number of characteristics setting them apart from the members of other classes. Classification of vegetation is not new, though until recently its basis has been largely subjective. *Ordination* is a more recent development, though one which has its roots in much of the earlier work on vegetation. It stems from the concept of vegetation as a continuum, which was discussed in the previous chapter. In ordination of stands an attempt is made to place each stand in relation to one or more axes in such a way that a statement of its position relative to the axes conveys the maximum information about its composition.

The two approaches are, in theory, quite distinct and based on fundamentally different concepts of the nature of plant communities. Classification implies discontinuity in composition not only between concrete units in the field, but also between abstract classes into which all vegetation may, theoretically, be placed. Ordination implies continuous variation in composition, though not precluding discontinuity in the field corresponding to discontinuity in determining factors. In practice the divergence is not so great as might appear. The reasons for this may be clarified by considering the nature of the variables used as criteria of classification. All biological classification is, to some extent, an arbitrary process. No two individuals are identical and in practice we must select certain features on which to base a classification and ignore others. Thus the taxonomist of flowering plants attaches great significance to features of carpel form and arrangement, and relatively little to stature and colour. Which features are regarded as important and which are not is ultimately determined by which lead to the most natural classification, i.e. which show the greatest amount of correlation with other features. The taxonomy of flowering plants is facilitated by the fact that many of the characters used are discontinuous, e.g. with very few exceptions carpels are either fused or free, flower parts are either whorled or spiral, etc.

The difficulties of classification would be many times increased if each character could only be described as a value for a continuous variable. Vegetation stands differ from one another primarily in terms of species composition, which represents a set of discontinuous variables—each species may be either present or absent. To this extent the problem of classification is, theoretically at least, fairly straightforward. There might be as many classes as possible combinations of species and no more. If any large proportion of these classes is actually found, a classification on this basis, though valid, would be of little practical value. Such a classification is unsatisfactory in two ways. It will certainly set apart widely different stands with no or few species in common and will distinguish groups of stands between which few or no intermediates are found because of the widely different tolerances of the constituent species. This is, however, merely formalizing common knowledge—it requires no system of classification to make clear the distinction between the vegetation of a sand dune and that of a raised bog. On a more detailed scale it is found that if all constituent species are included as contributory to the classification, then within stands composed of species of similar tolerances all possible combinations may be found, and the classification becomes so fragmented as to be valueless. On the other hand, if classification is based only on a selection of species, many classes will be too wide, including a variety of types evident even on casual inspection. This follows because the more prominent species in each stand, which are the only ones that can logically be selected, are normally those of wider range of tolerance.* Classification of stands on a basis of presence or absence of species tends to give either too broad or too narrow a classification for practical purposes. It is, therefore, necessary to consider the quantity of the more prominent species, with the result that classification becomes dependent on continuous variables.

* In opposition to this the Zürich–Montpellier school's dependence on faithful or characteristic species might be quoted, but their determination of faithful species rests on a circular argument. (See Poore, 1956.)

It is difficult to delimit a boundary between classes at some value of a single continuous variable but it is quite impracticable to do so on subjective criteria if a number of such variables have to be taken into account.*

Even if the arbitrary simplification is made of considering presence and absence of species only, the important difference between stands lies in the amount of different species. The accidental occurrence of one individual, surviving for a time in a stand in a habitat in which it normally does not occur, is not indicative of any special feature of that stand. We are, in fact, dealing with a population of individuals (if stands may be so regarded) which differ from one another in terms of continuous variables, of which presence and absence are only a crude expression. With only slight exaggeration we may, within the limits of a set of broadly similar stands, regard absence as simply the extreme value of a continuous variable.

Accepting that the features characterizing stands are continuous variables, there is still the possibility that the *total* characteristics of the stands may not vary continuously, i.e. that certain combinations of the abundance values of different species may occur, and others not, so that a natural boundary between one set of stands and another may be found and classes defined. Or, more likely, certain combinations may occur much more commonly than others. In this case these combinations will be more or less readily recognizable and more or less easily defined, but certain stands will be intermediate and not readily assignable to any class. This situation is very familiar in the field. There are certain clearly recognizable vegetation types which have commonly been named and described, and intermediates between them, occupying much smaller areas, generally styled ecotones or transition zones. The latter are often dismissed with a brief mention as being departures from the 'normal' pattern, a procedure which has tended to overemphasize the distinctiveness of units of classification. The same overemphasis has resulted from the insistence by the Zürich–Montpellier school on homogeneity of stands (in broad terms, not the strict sense) for description.

In practice then, classification, like ordination, often leads to the concept of vegetation as varying continuously in composition but it does emphasize the commoner occurrence of certain variants. Thus Poore (1955b, c), using a modification of the Zürich–Montpellier system, and comparing data for stands subjectively,

---

* There is an interesting parallelism, as Webb (1954) suggests, with the problem of classification of igneous rocks.

has demonstrated that the majority of stands do tend to fall into groups of similar composition, groups which he terms 'noda', with a relatively small proportion of intermediate stands. McVean and Ratcliffe (1962) in a regional account based on a classificatory approach found that 'as the work progressed it became increasingly apparent that variation in Highland vegetation is virtually continuous'. Various workers using a formal system of classification have found it expedient and informative to arrange their classes in an order showing the independent rise and fall of importance of different species (cf. Curtis, 1959, p. 481).

Although a classificatory approach may lead to the conclusion that vegetation is basically continuous, ordination is scarcely practical as a treatment of stands of vegetation differing widely in physiognomy. Thus Curtis (1959), in his account of the vegetation of Wisconsin, made an arbitrary classification into broad types which were then examined by ordinational techniques. It is at the lower levels of difference, where stands are comparable in physiognomy, that disagreement arises whether classification or ordination is more appropriate. Decision depends partly on the type of data available from stands; presence and absence data generally lead more readily to classification, quantitative data to ordination, but the alternative procedure is not impossible in either case, as consideration of the various techniques discussed below will show. The danger of classification is that the techniques involved may result in an over-emphasis of discontinuities. An ordination on the other hand, provided it does take cognizance of the greater part of the available information, will expose any discontinuities in composition that are present in the data. Unless, therefore, there are strong reasons arising from the nature of the data for preferring classification in a particular case, ordination is the sounder initial procedure even though the results of the ordination may conceivably indicate that a classification is the best means of summarizing the available information finally (Greig-Smith, 1961b). The objection sometimes raised against ordination, that it does not permit mapping of the vegetation, is not valid. An ordination may be divided up arbitrarily into units which may be mapped or otherwise used for descriptive purposes. The other principal objections advanced against ordination as a practical technique are the amount of data necessary from stands and the computational labour involved in their manipulation. Neither puts ordination at any great disadvantage against objective procedures of classification; the contrast is between subjective techniques (which do not permit of ordination in any real sense) and objective techniques of both classification and ordination.

Methods of sampling depend on the techniques of classification or ordination which it is intended to use. The stands examined may be either a random or systematic sample of those available, or may be selected to include a representative range of variation. In practice it is often more profitable to exercise some degree of conscious selection of stands. If this is done it must be remembered that a bias may be introduced in the selection and the resulting data may indicate a more clear-cut distinction into groups than actually exists. Most techniques are applicable to data including only a single value for each species in each stand, but some require replication of observations within stands and these may need to be randomly selected. It cannot be too strongly emphasized that careful consideration should be given to the choice of technique of analysis to be used before data are collected.

It is convenient to consider techniques of both classification and ordination under three heads: (1) those based on association between species, (2) those based on measures of similarity between stands, (3) those relating the composition of stands to environmental gradients. Closely related to the classification and ordination of stands is the problem of defining groups of species which show similar ecological behaviour. These four groups of techniques will be considered in turn.

The technique of analysis of stand data in terms of association between species depends on whether the species are recorded in terms of presence or absence or are assessed quantitatively. Goodall (1953a) examined presence and absence data from random quadrats for the occurence of interspecific association. The procedure is the same as was described for the examination of interspecific association in Chapter 4. The difference is in the size of quadrats used. Indeed, there is no fundamental distinction between the heterogeneity shown by a mosaic of relatively small-scale patches of different specific composition within what is commonly regarded as one community, and the larger-scale mosaic formed by patches of different communities. The interpretation of any particular case depends on further information about the vegetation, e.g. whether or not there is any evidence of dynamic relationship between the patches of different composition. If the quadrat used is large, it is reasonable to suppose that each quadrat throw will include all the phases of the community that have been regarded earlier as coming under the heading of pattern. In the case Goodall describes there is, perhaps, some doubt whether the quadrats used were large enough for this purpose. Moreover, he says that the whole area would be regarded by many ecologists as occupied by one association. However, his data illustrate the method very well and the question whether his final groupings represent different vegetation types or pattern within one type can be left open. His data from 256 quadrats showed a large number of cases of interspecific association. He suggests four possible procedures for classifying a number of

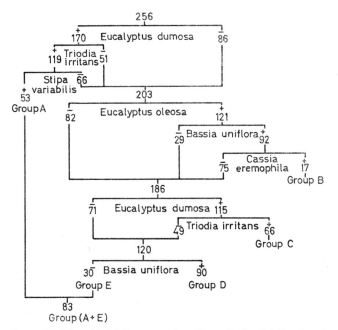

FIGURE 24. Victorian Mallee vegetation. Stages in the classification of quadrats into groups by taking as a potentially homogeneous group all quadrats containing one of two species showing positive association. At each stage the number of quadrats in which the species in question is present ($+$) and absent ($-$) is indicated (from Goodall, 1953a, by courtesy of *Aust. J. Bot.*)

quadrats showing association between two species into groups exhibiting no internal heterogeneity in species composition. These are to take as a potentially homogeneous group (1) all quadrats containing one of the two species, (2) all those not containing one of the two species, (3) all those containing both the two species, and (4) all those containing neither of the two species. After trial of all four methods Goodall concluded that the first was the most successful in eliminating association and also the least laborious. The course of the successive eliminations is shown in FIGURE 24. Of the species showing association, *Eucalyptus dumosa* was the most frequent and the 170 quadrats containing it were extracted first. These 170 quadrats were re-examined and among them *Triodia irritans* was the most frequent species showing association. 119 quadrats out of the 170 contained it and these 119 were re-examined, resulting in the elimination of 66 quadrats not containing *Stipa variabilis*. The 53 quadrats containing the latter showed no further associations and formed his homogeneous group A. The process was then repeated with the remaining 203 quadrats, resulting in the assignment of 17 quadrats to group B, and so on. Since the objective is to determine the minimum number of homogeneous groups, the groups are finally combined in pairs and each pair of groups examined for association, to test whether the recombination results in heterogeneity reappearing. In Goodall's case this resulted in groups A and E being combined. The four final groups could then be described in terms of frequencies of the species composing them (TABLE 17). The three other procedures did not result in all quadrats being assigned to the same groupings, but the definition of the groups in terms of specific frequency was closely similar, and 109 of the 256 quadrats were similarly allotted by all procedures. Thus, though the assignment of an individual quadrat to a group is to some extent arbitrary according to the exact procedure used, a clear demonstration of the existence of groupings, and a definition of them, is provided by examination of interspecific association.

Goodall's procedure is based on the formation of groups of stands by elimination of association. It is also characterized by dependence on positive associations, negative associations being ignored. Hopkins (1957), Agnew (1957, 1961) and McIntosh (1962) have concentrated rather on the use of both positive and negative associations to extract groups of species linked by positive associations. They are thus concerned rather with ecological groupings of species (considered further below) than with classification of stands, but they go on to consider stands in the light of these groupings. Hopkins equates a species grouping with a vegetational 'basic unit' and assigns quadrats to one or other 'basic unit' or even partly to one unit and partly to one or more others. Assignment of a quadrat to a 'basic unit' is determined by an arbitrary formula fixing the number of species, of those making up the unit, required to be present. Agnew uses the presence, and McIntosh uses the relative importance, of species of the different groupings to produce an ordination of the stands sampled. Thus in all three procedures arrangement of stands is derived indirectly.

CLASSIFICATION AND ORDINATION

TABLE 17

VICTORIAN MALLEE VEGETATION

FREQUENCY (PER CENT) OF SPECIES IN FOUR GROUPS OF QUADRATS BASED ON
PRESENCE OF SINGLE SPECIES SHOWING POSITIVE CORRELATIONS
Frequencies of 50 per cent or more have been given in bold type
(from Goodall, 1953a, by courtesy of *Aust. J. Bot.*)

Species	Group			
	A+E	B	C	D
	No. of quadrats			
	83	17	66	90
Acacia rigens A. Cunn.	2	0	5	0
Bassia parviflora Anders	2	0	0	4
B. uniflora R. Br.	17	**100**	5	**100**
Beyeria opaca F. Muell.	1	6	5	0
Callitris verrucosa R. Br.	1	0	26	0
Cassia eremophila A. Cunn.	1	**100**	0	2
*Chenopodium pseudomicrophyllum** (1)	5	6	2	22
*C. pseudomicrophyllum** (2)	12	6	0	13
Danthonia semiannularis R. Br.	20	0	6	7
Dodonaea bursariifolia Behr	**53**	0	**71**	17
D. stenozyga F. Muell.	1	0	0	6
Enchylaena tomentosa R. Br.	4	24	0	3
Eucalyptus calycogona Turcz.	14	18	11	30
E. dumosa A. Cunn.	**81**	12	**100**	39
E. oleosa F. Muell.	49	**100**	24	**81**
Grevillea huegelii Meiss	5	35	8	4
Kochia pentatropis R. Tate	2	24	0	7
Lepidosperma viscidum R. Br.	1	0	9	0
Lomandra leucocephala R. Br.	4	0	8	0
Melaleuca pubescens Schau.	4	0	8	0
M. uncinata R. Br.	**55**	0	**53**	3
Micromyrtus ciliatus J. M. Black	0	0	9	0
Myoporum platycarpum R. Br.	2	0	5	0
Olearia muelleri Bth.	5	**59**	0	6
Santalum murrayanum F. Muell.	1	12	5	1
Stipa elegantissima Lab.	1	12	0	3
S. mollis R. Br.	7	0	5	10
S. variabilis Hughes	**87**	**88**	0	**78**
Triodia irritans R. Br.	**78**	6	**100**	3
Vittadinia triloba D.C.	46	24	11	37
Westringia rigida R. Br.	6	**71**	5	3
Zygophyllum apiculatum F. Muell.	4	**65**	2	47

* *C. pseudomicrophyllum* Aellen occurs in this area in two forms.

165

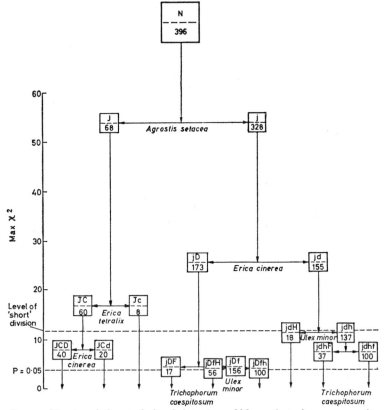

FIGURE 25. Association-analysis of data from 396 quadrats in a community containing ten species. Hierarchical analysis (from Williams and Lambert, 1960, by courtesy of *J. Ecol.*)

Williams and Lambert (1959, 1960; see also Williams and Lance, 1958) have examined more critically the problem of classification of stands for which only presence and absence data are available. They point out that Goodall's definition of 'homogeneous groupings' involves two independent concepts; significant associations may be absent in a grouping either by being indeterminate (one or both species being present or absent in all samples) or by associations falling below significance. These are not of equal importance, a non-significant association having been tested and rejected whereas an indeterminate association cannot be tested. They therefore prefer, if the choice has to be made, to reduce the level of association rather than render it indeterminate.

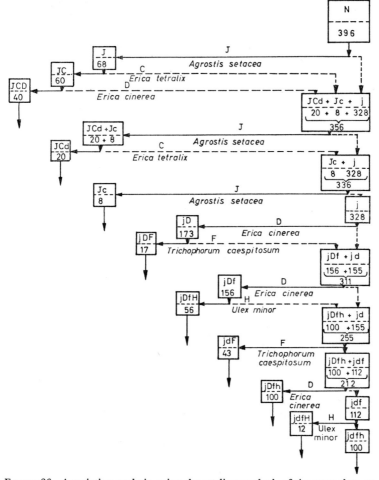

FIGURE 26. Association-analysis, using the pooling method, of the same data as in the analysis shown in FIGURE 25 (from Williams and Lambert, 1960, by courtesy of *J. Ecol.*)

The objective is so to sub-divide the population of stands that all associations disappear. Williams and Lambert point out that in general there will be a number of alternative subdivisions which will achieve this. They therefore introduce the concept of *efficient* subdivision, i.e. subdivision based on that species which produces the smallest total number of significant associations in the two sub-classes. For any but the simplest sets of data it is not practicable, even with electronic computing, to test the associations in the sets of subclasses that result from subdivision on all possible species. It is therefore necessary to choose some parameter I as a measure of association. Williams and Lambert show that, on theoretical grounds, division on the species with the maximum ΣI will tend to reduce the residual ΣI in the resulting subclasses to a minimum, i.e. to give efficient subdivision. Goodall's division on the most abundant of those species showing significant positive association will not necessarily do so.

Various parameters might be used as the index I. There are evident advantages in using χ^2 or some derivative of it, because χ^2 must be calculated as a test of significance. Having considered χ^2 uncorrected, χ^2 with Yate's correction, χ^2/N and $\sqrt{(\chi^2/N)}$, Williams and Lambert selected $\sqrt{(\chi^2/N)}$ (the correlation coefficient for presence and absence data). Subsequent theoretical consideration has shown that χ^2/N carries the maximum information and is therefore to be preferred to $\sqrt{(\chi^2/N)}$, but no practical test has yet been made (Williams and Dale, unpublished).

Goodall, having derived a homogeneous group of stands, pooled all the residual stands before re-examining associations. The alternative is to adopt a hierarchical procedure, both the groups resulting from a division being retained (FIGURE 25). Williams and Lambert prefer hierarchical division on the theoretical grounds that pooling is likely to give substantially the same final groups but by a longer route and that the final groups will be difficult to characterize and the route by which they are obtained will be meaningless. These objections to pooling are substantiated by a practical test (FIGURES 25 and 26).

Since Goodall's procedure involved 'pooling', the problem of significance arose only in connection with termination of subdivision. If division is hierarchical it is clear that the relative importance of different subdivisions of the same order may vary. Thus one subdivision may reduce the residual heterogeneity greatly while a corresponding division on the other side of the hierarchy may result in a relatively small decrease. Some measure of the heterogeneity of any class under examination is therefore needed as a means of assessing the fall in level of heterogeneity produced by a subdivision. After consideration of various other possible measures, which need not be elaborated here, Williams and Lambert conclude that the highest individual χ^2 within a class is the most suitable measure. This may conveniently be shown in the diagram representing the course of division, and provides a ready indication of the fall in relative heterogeneity produced at different stages. Thus in

Figure 25 the initial division reduces the highest single χ^2 from *ca.* 54 to *ca.* 14 in the '$\mathcal{J}$' (*Agrostis* present) class and to *ca.* 26 in the '*j*' (*Agrostis* absent) class.

The simplest criterion for termination of subdivision is to continue until no individual χ^2 exceeds 3·84 (corresponding to $P = 0·05$), giving what Williams and Lambert have termed 'long division'. They point out, however, that if a population has been divided down to this level and each quadrat is then replicated n times, the hierarchy will remain the same but all χ^2 values will be multiplied by n, so that further subdivisions, not previously regarded as significant, will be included at the lower end. Since it is generally the major divisions which are of interest, they suggest termination of subdivision at some level proportional to the total number of stands (N) being analysed. They have adopted the arbitrary termination value for χ^2 of $N·2^{-5}$ for this 'short division', as their experience shows that it will expose at least the majority of subdivisions which are susceptible of ecological interpretation.

This technique of 'association-analysis' has been programmed for electronic computation (Williams and Lambert, 1960) and applied to several sets of data from natural vegetation. It has not only produced subdivisions expected from previous knowledge of the areas, but has also exposed differences, previously overlooked, which are capable of ecological interpretation. The effect of size of sampling unit on indications of association has been discussed in Chapter 4 and is clearly relevant to association-analysis, as Lambert and Williams (1962) have appreciated. Use of small sampling units may introduce strong negative associations due to interspecific competition or even to spatial exclusion. Kershaw (1961) has demonstrated this by two association-analyses of data from a line of point quadrats spaced 1 cm apart in chalk grassland. The line was considered as consecutive segments, 20 cm and 40 cm long in the first and second analyses respectively. The course of subdivision was markedly different in the two analyses. His samples, taken with a covariance analysis of pattern also in mind, are, however, clearly too small to be satisfactory for association-analysis. Williams and Lambert used quadrats 1 m square, representing very much larger samples than the 20 or 40 points in Kershaw's data, and it is probable, in the types of vegetation with which they were concerned, that this was sufficiently large to eliminate features due to pattern within communities.

Association-analysis results in a hierarchical classification. Williams and Lambert emphasize that this hierarchical nature makes for readier interpretation in relation to the habitat. This is undoubtedly true for a particular set of samples; nevertheless, more importance is probably to be attached to the final groupings than to the course of subdivision by which they are obtained, because the route by which a particular grouping is segregated will depend on what other groupings are included in the samples. This may be made clear by considering a much simplified hypothetical case. Suppose a set of samples are taken from an area including two vegetation types, one composed of species *A* and *B*, the other of species *A* and *C*. The first (and only) division will be on *B* or *C*. If now another area including the *AB* type and also a type composed of species *B* and *D*, is sampled the division will be on *A* or *D*. Thus, the final *AB* grouping will appear in both cases but the species emphasized in its segregation will be different. This limitation applies to any hierarchical procedure applied to a limited number of the total entities; it is emphasized here because it has frequently been overlooked in classifications, both subjective and objective, of the vegetation of limited areas.

A comparable difficulty may arise if the method is applied to samples including widely diverse types as, for example, in primary survey of all the vegetation of a little known region. Two vegetation types with no species in common could segregate on any one of a large number of species; which one would emerge as the criterion might turn on the amount of additional association, perhaps representing little more than within community pattern, shown by the different species.

Association-analysis is concerned with qualitative data (species present or absent). Quantitative data, involving some measure of the amount of different species in the various stands, require a different approach. The obvious criterion of association between species is the correlation coefficient. The possibility of using correlation coefficients between amounts of different species in studies of their interrelationships and grouping has interested ecologists for some time. Stewart and Keller (1936) examined the correlation of cover of different species in various vegetation 'types' (defined in terms of the most abundant species present) in the semidesert shrub formation of Utah. Various conclusions were drawn about species relationships but Stewart and Keller did not attempt to base any formal grouping of stands on the results. Tuomikoski (1942) examined the correlation between frequencies of species in 1 m square quadrats of moorland in Finland and attempted to arrange species into groups the members of which showed mutual positive correlation but negative correlation with those of other groups. He was thus concerned rather with the detection of ecological groupings of species (see below p. 197 *et. seq.*) than with classification of stands.

Several attempts have more recently been made to apply the techniques of factor analysis, originally developed in relation to the assessment of psychological data, to the ordination of vegetation. Factor analysis has been applied by Goodall (1954c) to the same samples of mallee as he examined for species association, by Hamming (see De Vries, 1954b) to grasslands and by Dagnelie (1960) to beechwood, and to grassland (data of Ellenberg, 1956) and Wisconsin forest (data of Bray and Curtis, 1957).

The representation of any one species in a stand may be regarded as determined by the level of a number of factors.* Of these factors some will influence several or even all of the species present, others may be specific to individual species. If a pair of species were determined entirely by the same *common factors* they would show perfect correlation in their occurrence. The expected representation of a species j in a stand i will be

$$z_{ji} = a_{j1}F_{1i} + a_{j2}F_{2i} + a_{j3}F_{3i} + \ldots a_{jm}F_{mi} + a_j S_{ji}$$

where F_{1i}, F_{2i} etc. are the values of the common factors in stand i, S_{ji} is the value of the factor specific to species j and a_{j1}, and a_{j2} etc. are coefficients characteristic of the species j and reflecting the degree to which it is determined by each of the common factors. The resemblance of this expression to a multiple regression equation is evident. The difference lies in the fact that the values of F_{1j}, F_{2j} etc. are not known, nor indeed, are the factors themselves normally identifiable at this stage, if at all.

Provided all data are expressed in standard units (i.e. in terms of standard deviation from the mean for each species) and provided the values of the factors are independent of one another, it can be shown that the correlation coefficient between two species j and k is

$$r_{jk} = a_{j1}a_{k1} + a_{j2}a_{k2} + a_{j3}a_{k3} + \ldots a_{jm}a_{km}$$

Of the terms in this equation and the one above the z_{ji} are known and the correlation coefficients r_{jk} between all pairs of species can be calculated. The problem is to obtain the best estimate of the coefficients a, the 'loadings' of the species on the different factors. It is implicit in the approach that the whole matrix of correlation coefficients can be accounted for by substantially fewer factors than the number of species.

---

* The concept of 'factor', as an expression of correlation between the variables being examined, used in factor analysis must be kept clearly distinct from that of factor as commonly used in ecological contexts (see p. 18).

A useful short and clear exposition of the principles of factor analysis, discussed in a biological context, has been given by Ferrari *et al.* (1957).

No account of the computation involved will be given here; reference may be made to the various general accounts of factor analysis (e.g. Holzinger and Harman, 1941; Thurstone, 1947; Cattell, 1952; Kendall, 1957). There are, however, certain general considerations which should be discussed. The first step in analysis is the preparation of the matrix of correlation coefficients. The principal diagonal will contain the correlation coefficients r_{jj}, i.e. the correlation of the species j with itself. As calculated from the data this will naturally have the value 1. The correlation coefficient r_{jj} will, however, include not only items of the form a^2_{j1}, a^2_{j2} etc. (the squares of loadings on the common factors) but also an item a^2_{js} where a_{js} is the loading on the factor specific to that species, and, theoretically, though in practice not distinguishable from the last term, an item due to random elements in the representation of the species. There is a difference of opinion whether the analysis should proceed with unities in the principal diagonal of the correlation matrix (component analysis) or whether the unit values should be replaced by communalities, values representing the determination of the species by common factors (factor analysis in the strict sense). (The communalities can be estimated from the remainder of the matrix.)

Decision on the use of unities or communalities depends, at least in part, on the nature of the problem under investigation. Thurstone (1947) states that if the immediate object of the analysis is to reproduce the 'test scores' (here the values for species in each stand) as closely as possible, then unity should be written in the cells of the principal diagonal. If, on the other hand, the object is to reproduce the intercorrelations (between species), communalities should be used. Thus it would appear that, in the context of vegetation, if ordination *per se* is the primary object, unities should be used, but if interest centres on the relationship between species and their control by ecological factors, then communalities should be used. Bearing in mind that factor analysis is essentially an exploratory technique for use in simplifying a complex set of data where the controlling ecological factors are partly or entirely unknown, the principal objective is likely to be the clarification of the interrelationships and hence, generally, the use of communalities is to be preferred*. In practice, as will be shown below, it appears likely that the conclusions drawn will not differ substantially.

---

* W. T. Williams (personal communication) has pointed out that this preference for the use of communalities applies only if the objects being correlated are fairly similar, i.e. if the correlation matrix is substantially positive, and that difficulties will arise if the population is very discontinuous.

Various procedures for calculation of the loadings of species on the different factors have been devised. These will not be considered here. There is general agreement that the method of 'principal axes' is the most satisfactory. The others commonly used are approximations to this, designed to lessen the amount of computation involved. The loadings are commonly presented in the form of a table such as TABLE 18, which shows the first three factors extracted by Goodall from his data for mallee. The first factor, A, is that which will account for the maximum amount of correlation. Its extraction is followed by the calculation of a new matrix containing the residual correlation, from which a second factor is extracted,

TABLE 18

VICTORIAN MALLEE VEGETATION. LOADINGS ON THE FIRST THREE FACTORS EXTRACTED BY GOODALL (1954c, by courtesy of *Aust. J. Bot.*) USING THE PRINCIPAL AXES METHOD FOR UNITIES. VALUES OF h^2 FROM DAGNELIE (1960, by courtesy of *Bull. Serv. Carte phytogéogr.*)

Species	Factor A	Factor B	Factor C	h^2
1. *Bassia uniflora*	0·84	—0·10	0·14	0·73
2. *Chenopodium pseudomicrophyllum* (prostrate form)	0·31	0·57	0·09	0·43
3. *C. pseudomicrophyllum* (erect form)	0·15	0·73	0·02	0·55
4. *Danthonia semiannularis*	—0·56	0·48	0·22	0·59
5. *Dodonaea bursariifolia*	—0·74	0·08	—0·22	0·60
6. *Eucalyptus calycogona*	0·50	0·43	—0·50	0·69
7. *Eucalyptus dumosa*	—0·53	0·53	—0·34	0·68
8. *Eucalyptus oleosa*	0·76	—0·15	0·18	0·63
9. *Melaleuca uncinata*	—0·67	—0·21	0·55	0·80
10. *Stipa variabilis*	0·34	0·72	0·21	0·68
11. *Triodia irritans*	—0·76	—0·06	—0·34	0·70
12. *Vittadinia triloba*	0·04	0·43	0·59	0·54
13. *Westringia rigida*	0·48	0·06	—0·51	0·49
14. *Zygophyllum apiculatum*	0·77	—0·22	0·02	0·64

the process being repeated until no significant correlation remains. The last column shows the value of h^2, the sum of the squares of the loadings on the different factors for one species. This value indicates the proportion of the variance of a species which is accounted for by the factors extracted, the remainder resulting from any further significant common factors, any factor specific to the species and a random element.

The relationships of the species can be represented in geometrical form. The loadings of a species on the various factors will give the coordinates of its position in 'factor space'. FIGURE 27 shows the 14 species used by Goodall plotted in relation to the first two factors extracted. The figure shows the corresponding diagrams for (*a*) principal axes method for unities, (*b*) centroid method for unities, (*c*) principal axes method for communalities, (*d*) centroid method for communalities. It will be seen that although the diagrams are all different, the relative placing of species remains comparatively constant. This suggests that the precise method of analysis does not

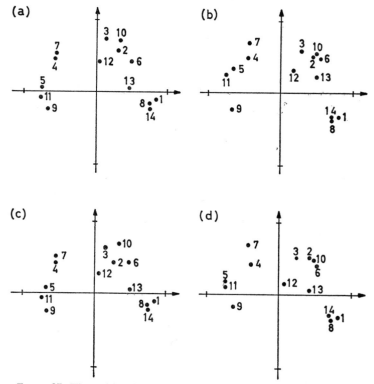

FIGURE 27. The positions in 'factor space' of the 14 species used in Goodall's (1954c) analysis of Victorian Mallee. (a) Principal axes method for unities; (b) centroid method for unities; (c) principal axes method for communalities; (d) centroid method for communalities (redrawn from Dagnelie, 1960, by courtesy of *Bull. Serv. Carte phytogéogr.*)

affect the ecological conclusions to a serious degree. Dagnelie (1960, p. 105) presents similar diagrams for his beechwood data, which also indicate that the relative spacing of the species is not markedly affected by the method of analysis.

At this stage of the analysis an ordination of species in factor space has been obtained. From this, conclusions of ecological interest can be drawn. Species close together in this ordination are similar in their behaviour in relation to the factors extracted and hence, in so far as these factors in turn reflect ecological factors, similar in their ecological responses; as Dagnelie points out, such groups of species will be ones of value as differential species. The distance of a species from the origin is h; thus, species near the origin are ones the occurrence of which is not closely linked to the values of the factors. These will include not only species associated mainly with a specific factor rather than common factors (indicator species), as Dagnelie points out, but also those with a largely random distribution, i.e. those for which the total range of conditions in the stands examined is well within the limits of tolerance for the species concerned. The ordination of the species has been illustrated in FIGURE 27 in relation to two axes only. There is clearly no theoretical limit to the number of axes defining the factor space. A three-dimensional model may be used to represent a three axis space; further increase in the number of axes makes graphical representation very difficult with corresponding difficulties in interpretation.

The axes corresponding to the extracted factors are used to derive the ordination of the species but the ordination itself is independent of them. It may prove helpful in interpretation to rotate the axes so that the first axis corresponds to an evidently meaningful ecological gradient, indicated, for example, by the occurrence at its opposite ends of groups of species of known, ecologically contrasting characteristics. Techniques of carrying out such a rotation or even of referring the ordination to oblique axes are available. The decision whether or not to change the axes must be made on biological grounds and in the light of knowledge of the species and vegetation involved.

Since estimates of the loadings of the various species on different factors can be made, and the representations of species in different stands are known, it is clearly possible to calculate estimates of the values of the factors for each stand. It is again inappropriate to discuss the techniques of doing so here, but it is clear that the result will be an ordination of stands in terms of the values of the factors calculated for each. This ordination, like that of species, can be represented graphically in terms of coordinates on two or more axes. (Note, however, that whereas in the species ordination there is an upper limit of 1 to the value on any axis, corresponding to complete dependence on that factor, there is no such limit in the stand ordination.) Interpretation of the stand ordination depends on independent information on the species concerned. Thus Goodall (1954c) showed that the species with the highest positive loadings on the first factor in his analysis of mallee were those found in hollows with poor drainage, while those with high negative loadings were those found on the intervening ridges. He deduced that the first factor represented broadly a soil moisture gradient. He interpreted the second factor as representing a non-linear component of response to soil moisture but the remaining three factors proved difficult to interpret. W. T. Williams and M. B. Dale (personal communication) have re-examined Goodall's data and consider that the difficulty of interpreting the factors results from discontinuity in the data (see p. 177, below). It should be emphasized that although it may be possible to interpret the extracted factors (and only if this is so is the analysis likely to be rewarding), the factors themselves are entirely phytosociological in nature, being expressions of correlation of occurrence between species and having no meaning apart from them. Extraction of factors has two valid functions, (1) the simplification of complex data, permitting their readier inspection and interpretation, and (2) (particularly when communalities are used and the axes rotated) the production of hypotheses which may then be tested against other data.

Finally, it should be noted that the techniques of factor analysis are not limited by the form of distribution of the original data. It is even possible to use presence and absence data, as Dagnelie (1960) has demonstrated for ecological data. Williams and Lambert's association-analysis is essentially an approximation to factor analysis based on presence and absence of species.

Factor analysis has been discussed above in terms of analyses of stands (or other vegetation samples) as individuals and of species as 'attributes' or 'tests' of the underlying factors, the basic data being correlations between all possible pairs of species as they occur in different stands. The various techniques used in this way have been designated 'R-techniques' (see Cattell, 1952). It is also possible to treat the species as individuals and the stands as tests, all possible pairs of stands being examined in terms of the correlation of representation of different species in the two stands ('Q-techniques'). Thus if species A, B, C, . . . have a representation in stands 1, 2, 3, . . . of a_1, a_2, a_3 . . . , b_1, b_2, b_3 . . . , c_1, c_2, c_3, . . . etc., R-techniques involve correlations between such sets of pairs of values as a_1, b_1; a_2, b_2; a_3, b_3; . . . (correlation between species A and B) and Q-techniques correlations between such sets as a_1, a_2; b_1, b_2; c_1, c_2; . . . (correlation between stands 1 and 2). Q-techniques are thus based essentially on estimates of similarity between stands and result directly in an ordination of stands in factor space. The correlation coefficient is often not appropriate to stand comparison and any one of the wide range of indices of similarity could be used. Dagnelie (1960) has applied Q-techniques both to his own data and to those of Ellenberg (1956) and Bray and Curtis (1957).

Once the initial matrix of correlations has been produced the procedure is the same as in R-techniques; the same choices arise of unities or communalities in the principal diagonal, and of method of calculation of loadings. Since Q-techniques give a stand ordination directly, and in most problems this is likely to be required, they have evident advantages. Moreover, both similarity coefficients and Q-techniques in general are statistically robust (Sneath and Sokal, 1962; Sneath in Williams and Lambert, 1961c), i.e. the results obtained are little affected by the similarity coefficient and technique of analysis used.

Factor analysis is thus a very flexible technique, adapted especially to exploratory investigations of complex multivariate data, and will certainly play an important part in plant ecology in the future. However, like any other technique, it has limitations and these must be considered. The most obvious disadvantage, the amount of computation involved, is likely to be of less importance as electronic computing facilities become increasingly available. Nevertheless, even with electronic computing, factor analysis of any extensive data makes heavy demands on machine time if not so much on the ecologist's time, and the number of species or stands, or both, that can be handled may be limited. If factor analysis is intended, it is therefore important that the collection of data should be planned with this technique in mind. There are two important assumptions inherent in factor analysis—(1) that the factors are independent of one another and (2) that the representation of a species is linearly related to the factors affecting it. Curtis (1959, p. 482) has pointed out that 'although various measurable attributes of plant communities are no doubt related to environmental factors, there is no assurance that these factors are themselves independent; hence there is no assurance that the unities' (i.e. factors) 'extracted by the analysis are real'. Similarly, there can be no certainty that control can be represented by a linear relationship; indeed, general ecological experience indicates that frequently it will not be. Only further experience of interpretation of analyses can determine how serious these limitations are. Goodall's difficulty in interpreting his third and later factors perhaps reflects these difficulties. Williams (in Williams and Lambert, 1961c) has pointed out that if there is complete discontinuity in the original data, although the first factor will separate the two entities, any further factors are likely to be unrelated for the two entities and subsequent factors extracted

177

from a single analysis may well be incapable of interpretation. Hence it is desirable to make completely separate analyses of the entities. This emphasizes that factor analysis is likely to be more valuable in analysing data from a group of similar stands than in broader investigations.

The techniques of factor analysis, as we have seen, can be applied both to data on species correlation and data on stand correlation or stand similarity. The possibilities of using an index of similarity in drawing up a classification of a set of stands have attracted attention for some time. Sørensen (1948) calculated values of the coefficient associated with his name (see p. 136) for all pairs of stands and placed in one group stands between which the value of the coefficient was at least 50 per cent. Groups defined in this way

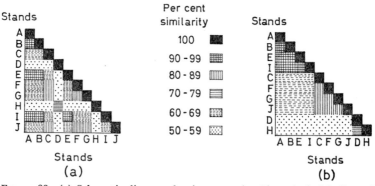

FIGURE 28. (a) Schematic diagram showing a matrix of hypothetical indices of similarity between pairs of stands. (b) The same indices arranged with similar stands adjacent to one another. Clearly distinct groups appear as triangles of high values. (Modified from Sneath and Sokal, 1962, by courtesy of *Nature, Lond.*)

were then associated into '2nd order groups' for which the limiting value of the coefficient was 40 per cent, and so on. In this way an objective classification was produced. Looman and Campbell (1960) have pointed out that the significance of a given value of the coefficient, as an indicator of degree of similarity, varies with the total number of species involved and the number in each of the stands being compared. They use the 0·1 per cent value of χ^2 to calculate the least number of species common to the two stands that will indicate association between the stands. Their detailed conclusions on Sørensen's results require modification as they used the χ^2 value for two degrees of freedom instead of that for one degree of freedom, appropriate to a 2×2 contingency table, but they indicate that Sørensen's arbitrary levels did not lead to gross misgrouping.

Such testing of the significance of each comparison in isolation provides only an approximate guide, as what is at issue is the significance of the whole matrix of similarities; no simple test exists for this.

Various workers have used indices of similarity as a guide to the relative placing of stands by inspection. One simple procedure is to rearrange the order of stands in the matrix of indices in such a way that high values are placed as near as possible to the diagonal (FIGURE 28). Clearly distinct groups will appear as triangles of high values as in FIGURE 28(b). Vegetation data are unlikely to

TABLE 19

UPLAND HARDWOOD FORESTS IN WISCONSIN
AVERAGE IMPORTANCE VALUE (IV) AND CONSTANCY % OF TREES IN STANDS
WITH GIVEN SPECIES AS THE LEADING DOMINANT
(from Curtis and McIntosh, 1951, by courtesy of *Ecology*)
(For species with highest importance potential only—80 stands)

Species	Leading dominant in stand			
	Q. velutina	*Q. alba*	*Q. rubra*	*A. saccharum*
Q. velutina				
Average IV	165·1	39·6	13·6	0
Constancy %	100·0	72·3	38·3	0
Q. alba				
Average IV	69·9	126·8	52·7	13·7
Constancy %	100·0	100·0	97·1	66·7
Q. rubra				
Average IV	3·6	39·2	152·3	37·2
Constancy %	25·0	94·5	100·0	76·3
A. saccharum				
Average IV	0	0·8	11·7	127·0
Constancy %	0	5·6	29·4	100·0

give as clear cut distinctions as this and whether the final arrangement is interpreted in terms of classification (e.g. Guinochet and Casal, 1957) or ordination (e.g. Clausen, 1957; Dix and Butler, 1960 etc.) appears to depend on the viewpoint preferred. Indices weighted by measures of representation of species have been used in similar ways. Thus Clausen examined her data in terms of frequency as well as of presence of species and Raabe (1952) used the mean constancy difference of species to assess the similarities between abstract communities.

A group of ordination techniques developed by Curtis and his associates, though not at first based directly on stand similarity, are most conveniently considered together. They are of particular interest because they are, so far, the only ones which have been extensively used on a wide range of vegetation types. The results have been described in Curtis' book on the vegetation of Wisconsin (Curtis, 1959) and in various subsequent papers.

A technique for obtaining a linear ordination of stands was first

TABLE 20

UPLAND HARDWOOD FORESTS IN WISCONSIN
AVERAGE IMPORTANCE VALUE (IV) AND CONSTANCY % OF TREES IN STANDS
WITH GIVEN SPECIES AS THE LEADING DOMINANT
(from Curtis and McIntosh, 1951, by courtesy of *Ecology*)
(Eleven species of intermediate importance potential—80 stands)

Species		Leading dominant in stand			
		Q. velutina	*Q. alba*	*Q. rubra*	*A. saccharum*
Q. macrocarpa	IV	15·6	3·5	4·2	0·1
	C %	50·0	38·9	20·6	4·8
Prunus serotina	IV	21·4	21·8	5·9	1·4
	C %	87·5	89·0	64·8	19·0
Carya ovata	IV	0·3	8·8	5·2	5·9
	C %	12·5	61·2	38·3	33·3
Juglans nigra	IV	1·5	1·2	2·2	1·9
	C %	12·5	11·1	20·6	23·8
Acer rubrum	IV	3·9	2·3	2·4	1·0
	C %	12·5	33·3	23·5	4·8
Juglans cinerea	IV	0	2·7	1·7	4·8
	C %	0	11·1	20·6	47·6
Fraxinus americana	IV	0	1·9	5·1	7·6
	C %	0	11·1	20·6	42·8
Ulmus rubra	IV	4·6	7·7	8·3	32·5
	C %	25·0	27·8	53·3	85·7
Tilia americana	IV	0·3	5·9	19·0	33·0
	C %	12·5	16·7	73·5	100·0
Carya cordiformis	IV	2·5	5·8	4·1	8·2
	C %	12·5	33·3	41·2	66·7
Ostrya virginiana	IV	0	2·4	5·5	16·2
	C %	0	22·2	41·2	95·3

developed. This may be illustrated from the account of upland hardwood forests given by Curtis and McIntosh (see also Brown and Curtis, 1952). Ninety-five stands of forest, chosen as objectively as possible, were examined; all known stands which satisfied the requirements of natural origin, adequate size, freedom from interference, and situation on upland land forms were included in the survey. The stands, after sampling by a random pairs method, were described in terms of Curtis's importance value (p. 142 above). Only a few species of trees ever had a high importance value. In 80 out of the 95 stands the 'leading dominant' (species with the greatest importance value) was one or other of four species. If the occurrence, in terms of average importance value and constancy, of these four species in the four groups of stands specified by the four leading dominants is examined (TABLE 19), it will be seen that if the leading

TABLE 21

UPLAND HARDWOOD FORESTS IN WISCONSIN
TREE SPECIES FOUND IN STANDS STUDIED, WITH THE
CLIMAX ADAPTATION NUMBERS OF EACH
(from Curtis and McIntosh, 1951, by courtesy of *Ecology*)

	Climax adaptation number
Quercus macrocarpa Michx.	1·0
Populus tremuloides Michx.	1·0
Acer negundo L.	1·0
Populus grandidentata Michx.	1·5
Quercus velutina Lam.	2·0
Carya ovata (Mill.) K. Koch.	3·5
Prunus serotina Ehrh.	3·5
Quercus alba L.	4·0
Juglans nigra L.	5·0
Quercus rubra L.	6·0
Juglans cinerea L.	7·0
Ulmus thomasi Sarg.	7·0
Acer rubrum L.	7·0
Fraxinus americana L.	7·5
Gymnocladus dioica (L.) Koch.	7·5
Tilia americana L.	8·0
Ulmus rubra Muhl.	8·0
Carpinus caroliniana Walt.	8·0
Celtis occidentalis L.	8·0
Carya cordiformis (Wang) K. Koch.	8·5
Ostrya virginiana (Mill.) K. Koch.	9·0
Acer saccharum Marsh.	10·0

* Climax adaptation number of these species is tentative, because of their low frequency of occurrence in this study.

dominants are arranged in the order *Quercus velutina*, *Q. alba*, *Q. rubra*, *Acer saccharum*, each species rises to a maximum importance value and constancy in the group of which it is the leading dominant and then falls off again. The occurrence of trees of lower importance value (TABLE 20) and of a selection of understorey species, shows the same regularity, with minor exceptions. The groups of stands were next subdivided by taking into account the species showing the second highest importance value. Thus stands with *Q. alba* the

181

leading dominant and *Q. velutina* the second dominant were placed nearer the *Q. velutina* end of the series than those with *Q. alba* the leading dominant and *Q. rubra* the second dominant. In this way the tree species could be arranged in an order (based on the position of their greatest average importance value in relation to the series) from those most nearly associated with *Q. velutina* to those most nearly associated with *A. saccharum* (TABLE 21). Other evidence suggests that the former represent pioneer and the latter climax species.

Having established the existence of this continuum of specific composition, Curtis and McIntosh suggested a method of placing individual stands in sequence in it. Each species was assigned a 'climax adaptation' number ranging from 10 for *Acer saccharum* to 1 for species at the other extreme (TABLE 21). This was necessarily somewhat arbitrary; while the species were given numbers in order of position along the series, it was bound to be a matter of opinion whether two species with similar, but not identical, occurrence in relation to the leading dominants should be assigned the same values, or values differing by 0·5, the smallest unit used. To assess the position of a single stand the importance values of the species present in the stand were weighted by their 'adaptation' numbers and added to give a figure termed the *continuum index*. Since the total importance value for a stand is always 300, being composed of total relative frequency plus total relative density plus total relative 'dominance', the continuum index ranges from 300 for a stand composed entirely of species having 'climax adaption' number 1 to 3000 for a stand composed entirely of *Acer saccharum*. The continuum index thus utilizes information from all species present in determining the position of a stand in the series, the species contributing to the determination according to their importance value. If stands were arranged in order of their continuum indices, the importance value of species was found to rise to a maximum and then fall off, each species having a characteristic position of the peak in relation to the continuum index (FIGURES 29 and 30), thus confirming the continuous nature of variation in composition of the stands. The distribution of the understorey species has been more extensively investigated by Gilbert and Curtis (1953) and that of the soil microfungi has been examined by Tresner *et al.* (1954). The species of both groups were found to show a definite pattern in relation to values of the continuum index.

This approach is clearly a useful one, provided a bias is not introduced in choice of stands for inclusion. It is inevitable that a certain subjective element must enter into their selection, as Curtis

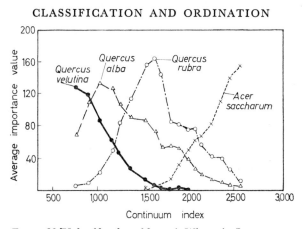

FIGURE 29. Upland hardwood forests in Wisconsin. Importance values of four major species arranged in numerical order of continuum index and averaged by successive groups of 5 stands (from Curtis and McIntosh, 1951, by courtesy of *Ecology*)

and McIntosh recognize, but this may be kept to a minimum, either by including all available stands satisfying certain conditions, as was done in this case, or, if the number of such stands is too great, or if a continuous area is being examined, by selecting the sampling sites randomly. Provided that the selection of stands is as nearly unbiased as possible and the sampling procedure is satisfactory, the arrangement of species in a series from 'pioneer' to 'climax' is objective.

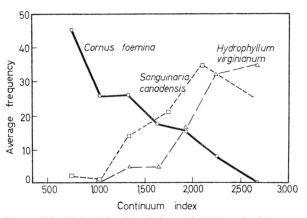

FIGURE 30. Upland hardwood forests in Wisconsin. Average frequency values of three herbs, arranged in order of continuum index (from Curtis and McIntosh, 1951, by courtesy of *Ecology*)

The use of importance value in the description of the stands is not an essential part of the method; any quantitative measure might be used. It is, perhaps, not even essential that it should be put on a relative basis, though probably it would be better to do so. A situation might be imagined where, for example, two species of the same climax adaptation number might both have 100 per cent frequency or nearly so when growing together. If absolute values were used a community of both species would have a continuum index twice that of either species in a pure stand, though whether one or both were present might be due to chance factors of dispersal in identical habitats. An arbitrary element is introduced in the assignment of 'climax adaptation' numbers. There is no certainty that a difference of one unit in one part of the series represents the same degree of difference in 'adaptation' as in another part of the same series and still less in another continuum. This objection is not, however, serious as, though the form of the curve of importance values for a species may be distorted, the relative positions of the curves of different species and especially of their maxima will not.

Curtis and McIntosh weighted each species by a value, the 'climax adaptation number', representing the position of its maximum development along the principal gradient of composition. Dix (1959), in a study of the effect of grazing on prairies in Wisconsin, has shown that species may be weighted according to their response to a particular environmental factor and stands thereby ordinated in relation to the incidence of that factor. By examination of a series of pairs of stands differing from each other only in occurrence or non-occurrence of grazing, Dix obtained values for the sum of densities of a species in grazed and ungrazed stands. The difference, expressed as a proportion of the sum of densities in ungrazed stands (where the difference was positive) or grazed stands (where the difference was negative), was used as an index of grazing susceptibility for the species. These indices were then used, in the same manner as the climax adaptation numbers, to produce an ordination of stands. Such an ordination will not necessarily account for the maximum amount of variation between stands but rather for that variation which reflects the incidence of grazing.

Though there is some choice of axis extracted by this technique, depending on the basis of weighting used, it suffers from the limitation that it can take cognizance of one axis of variability only; stands placed close together may in fact differ considerably in composition. Bray and Curtis (1957) have developed a procedure, based on measures of similarity between stands, which is not limited in this way. This procedure was elaborated in relation to the same vegetation type, upland hardwood forest of Southern Wisconsin, as that studied by Curtis and McIntosh, permitting comparisons between the results of the two treatments.

A total of 59 stands was examined, and described in terms of absolute number and basal area per acre of tree species and of frequency of shrub and herb species in 1 m square quadrats. The ordination was based on the 12 tree species with the highest presence values and 14 shrub and herb species; the latter were chosen at random within blocks along the axis produced by Curtis and Mc-Intosh, from among those species which were neither markedly rare nor markedly common. Measures of density and basal area were included separately for tree species, so that 38 items contributed to the specification of stands. The scores for a particular species were first expressed as percentages of the maximum recorded for that species. The scores for the various species in each stand were then placed on a relative basis. The index of similarity used was the coefficient of community, $\dfrac{2w}{a+b}$ (w, the sum of the lower score for each species; a, the sum of scores for one stand; b, the sum of scores for the other stand). When scores are on a relative basis this reduces to w.

The stands were arranged in such a way that the distances between them were proportional to the differences in their composition. Clausen (1957) attempted to arrange stands in a two-dimensional framework by inspection but a more precise method is clearly desirable. Bray and Curtis used a geometrical method which may be explained most readily from a hypothetical example. TABLE 22 shows, in the upper right-hand portion, the coefficient of community for five stands and, in the lower left portion, the corresponding measure of difference, or spatial separation (100 minus the coefficient of community). The maximum distance between stands is 99·9 between stands 1 and 2, which are therefore placed at opposite ends of the first (x) axis [FIGURE 31(a)]. Stand 3 is distant 70 units from both stands 1 and 2. Its position along the x axis is thus determined by the intersection of arcs of radius 70 units centred at points 1 and 2. Similarly, stand 4 is located by the intersection of arcs of radius 70 units centred at point 1 and radius 50 units centred at point 2. Placing in relation to the second (y) axis is obtained by using as reference points two stands close together on the x axis, but separated by a relatively large interstand distance. Here stands 3 and 4 are used [FIGURE 31(b)]. That they are on opposite sides of the x axis is indicated by their interstand distance; which is placed above and which below the x axis is decided arbitrarily.

TABLE 22

MATRIX OF HYPOTHETICAL EXACT INTERPOINT
DISTANCES

(from Bray and Curtis, 1957, by courtesy of *Ecol.
Monogr.*)

Stand No.	1	2	3	4	5
1	—	0.1	30	30	30
2	99·9	—	30	50	50
3	70	70	—	17.8	79.6
4	70	50	82.2	—	35.2
5	70	50	20.4	64.8	—

The upper right portion of the table shows hypothetical data on point similarity for an exact spatial system. The lower left portion shows data on similarity, inverted to give interpoint distances.

Which of the two possible positions for stand 5 is correct is determined by its distances from points 1 and 2. Beals (1960) has pointed out that the distance along the first axis at which a stand, e.g. stand 3, is placed is readily obtained by solving the two right-angled triangles formed by, in this case, the lines 13, 23, 12 and the perpendicular from 3 on to 12. In this hypothetical example the interstand distances can be satisfied by placing the stands in a two-dimensional frame. If necessary, however, a third axis may be used, based on points with similar *x* and *y* locations but relatively large interstand distance. Further axes may be used, though graphical representation then becomes impossible.

Coefficients of community calculated from field data will be subject to a sampling error, i.e. replicate samples from the same

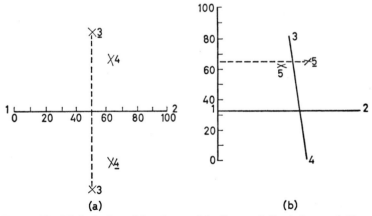

FIGURE 31. Method of stand location used by Bray and Curtis (see text) (from Bray and Curtis, 1957, by courtesy of *Ecol. Monogr.*)

stand will give a coefficient not of 1 but of a lower value. Bray and Curtis sampled two stands seven times and found a mean coefficient of 82 within each of the seven replications. A value of 80 was therefore considered the maximum coefficient (i.e. the value for two identical stands) and coefficients were subtracted from 80 instead of from 100 to obtain interstand distances. From the matrix of interstand distances the position of each stand relative to three

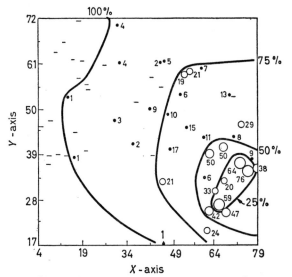

FIGURE 32. Upland hardwood forests in Wisconsin. Ordination of the stands examined by Bray and Curtis (1957), shown in relation to the x and y axes, and the basal area of *Tilia americana* as 100 in²/acre in each stand. Each symbol represents the location of a stand in the ordination. The largest circles represent stands in the upper 25 per cent of those occupied by *T. americana*, medium circles those in the 50–26 per cent quartile, small circles those in the 75–51 per cent quartile and dots those in the 100–76 per cent quartile. Dashes represent stands at which *T. americana* was absent. Contour lines are drawn to include all stands assigned to the indicated classes (from Bray and Curtis, 1957, by courtesy of *Ecol. Monogr.*)

axes was calculated in the manner outlined above. Examination of correlation between actual interstand distances in the resulting three-dimensional ordination and the observed coefficients of community showed that the ordination did account satisfactorily for the latter.

FIGURE 32 shows the position of stands in relation to the x and y axes, and the manner of plotting species occurrence on to the ordination. TABLE 23 shows the position of the midpoints (point of maximum representation) of the tree species on the three axes. Comparison with TABLE 21 shows that the species arrangement along the x axis corresponds closely with that in Curtis and McIntosh's linear ordination. This is not surprising since both procedures are designed to account for the most evident compositional gradient first. This table also brings out the additional information contained in the y and z positions; compare, for instance, *Juglans nigra* and *Ostrya virginiana* separated by only 1·1 units on the x axis but by 6·1 units on the y axis and 7·0 units on the z axis.

TABLE 23
UPLAND HARDWOOD FORESTS IN WISCONSIN
Location of species midpoints on the three axes of
Bray and Curtis' ordination (from Bray and Curtis,
1957, by courtesy of *Ecol. Monogr.*)

Species	X	Y	Z
Quercus macrocarpa	16·3	46·4	38·3
Quercus velutina	17·4	49·3	41·0
Carya ovata	30·1	44·8	29·5
Prunus serotina	31·5	46·5	30·1
Quercus alba	35.5	46.7	32.9
Quercus borealis	40·9	51·8	42·7
Ulmus americana	44·8	33·5	41·9
Populus grandidentata	47·6	45·4	49·4
Juglans nigra	56·6	41·0	43·4
Ostrya virginiana	57·7	34·9	50·4
Fraxinus americana	62·2	31·8	45·5
Juglans cinerea	63·1	40·4	43·7
Carya cordiformis	63.3	43.6	42.7
Tilia americana	64.9	37.5	49.9
Ulmus rubra	67.8	42.9	38.3
Acer saccharum	68.4	40.5	48.8

The labour of calculating the complete matrix of coefficients of community increases rapidly with increasing number of stands. Maycock and Curtis (1960) have suggested that if sufficient information is available to pick out the two most unlike stands, or two small groups of stands containing them, it is sufficient to prepare a partial matrix showing the coefficients with these stands only. They also suggest simplifying the calculation of position along the axis by using an arithmetic method instead of the geometric method of Bray and Curtis. In effect, they place a stand by taking the mean of the two positions indicated by the distance from one reference stand and by the total length of axis less the distance from the second reference stand.

Curtis (1959) has extended this approach to a consideration of all the major natural communities of Wisconsin. The similarity of communities was examined in terms of species of the ground layer, all species which were prevalent in any one community being taken into account (a total of 387 out of 867 species recorded). As with variation within one forest type, three axes were necessary and sufficient to accommodate the observed inter-community distances.

In addition to their extensive use in Wisconsin these techniques have been used in various other areas. Anderson (1960, 1963) has shown that the one-axis technique of Curtis and McIntosh can usefully be used to examine variation of a degree little greater than within-community pattern and Gittins (1963) has used the Bray and Curtis technique for a similar purpose. Of particular interest is Ashton's (1962) demonstration of the utility of the latter technique in the highly heterogeneous and species-rich rain forest of Brunei. Gimingham (1961), in an examination of heaths over a wide geographical range in western Europe, found that a one-axis ordination based on Curtis and McIntosh's technique was insufficient to account for the observed variation and, by inspection, placed stands on a network of axes. Falinski (1960) adopted a similar arrangement based on coefficients of similarities between stands.

The technique of Bray and Curtis has evident similarity to factor analysis by a Q-technique. Indeed, it might be regarded as a crude approximation to factor analysis. Dagnelie (1960) has subjected the same data to factor analysis, extracting three factors. The loadings on the second factor show strong correlation with position on Bray and Curtis' x axis ($r = + 0.9637$, P $\ll 0.001$). The loadings on the first factor are significantly though less strongly correlated with position on the y axis ($r = + 0.3789$, P $0.001–0.01$) but the loadings on the third factor show no relationship to position on the z axis ($r = - 0.0210$, P $0.8–0.9$). It is of some interest that the correlation of the first factor should be with the y axis rather than the x axis, but it must be remembered that it is the placing of stands relative to one another that is the objective of ordination; the axes used in construction of the ordination are of minor interest only. A more direct test of comparability is to examine the interstand distances in the two ordinations. A random sample of twenty-five interstand distances gave a correlation coefficient of $+ 0.8699$ (P $\ll 0.001$). The close agreement makes it doubtful whether the factor analysis has sufficient advantage over the cruder technique to justify the computational labour involved, at least over as wide a range of variation as is involved in this case.

An interesting comparison has also been made by Gittins (1963), who applied Bray and Curtis' technique and Williams and Lambert's association-analysis to the same data. Here, as one technique results in ordination and the other in hierarchical classification, no formal comparison of the two can be made. The ecological conclusions drawn were, however, substantially identical.

Harberd (1960, 1962), although adopting a classificatory approach, has interpreted his data in a manner essentially similar to a one-axis ordination. He has considered a modified form of Mahalanobis' D^2 as a measure of difference between stands each described by a single species list. Each stand is regarded as a unique sample of a hypothetical community, so that the intragroup variances and covariances of species normally involved in the calculation of D^2 are eliminated. Under these conditions D^2 between two stands reduces to the number of species represented in one only of the two stands. The absolute value of D^2 varies according to the total number of species present and Harberd therefore prefers to use 'relative D^2', expressed as a percentage of the total number of species present. In this form it is, as Harberd points out, the complement of Sørensen's coefficient, and is the same measure of distance as that of Bray and Curtis, but based on presence rather than amount. By associating lists for individual quadrats of *Agrostis–Festuca* grassland on this basis Harberd reduced eighty lists to seven groups, which were themselves arranged in such an order that each group was placed next to those most similar to it. Representation of individual species in the groups then showed consistent rise, consistent fall or unimodal distribution.

A number of techniques of ordination of stands have been outlined above. An ordination of stands is of some interest and value in itself, particularly in relatively little known vegetation. Of much greater value is its use as a means of elucidating the behaviour of individual species and the relation of vegetation to environmental factors. The ordination of stands provides a framework on which can be plotted any variable, for which a value can be assigned to individual stands. FIGURES 29 and 30 show examples of the distribution of individual species plotted on a one-axis ordination. FIGURE 32 shows one view of the distribution of a species on a three-axis ordination. It is on the examination and comparison of such individual species distributions that the distinctness or otherwise of communities must be judged. If the stands examined fall into a limited number of clearly defined communities, then there will be a correspondingly limited number of patterns of distribution of species in relation to the ordination. If, on the other hand, the vegetation varies continuously in composition each species may be expected

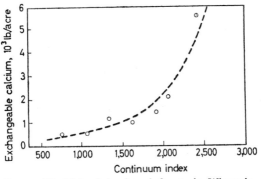

FIGURE 33. Upland hardwood forests in Wisconsin. Average level of exchangeable calcium in soil of stands arranged in order of continuum index (from Curtis and McIntosh, 1951, by courtesy of *Ecology*)

to have its own distinctive distribution. There is increasingly strong evidence from ordination studies that, in general, vegetation does vary continuously.

Environmental factors may be plotted on an ordination of stands in a similar way. FIGURE 33, from Curtis and McIntosh, shows an example of a soil factor plotted on a one-axis ordination. FIGURE 34 shows two soil factors plotted on the general ordination of Wisconsin communities produced by Curtis (1959). In each case there is

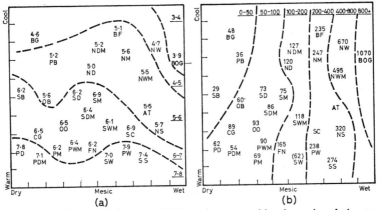

FIGURE 34. General ordination of Wisconsin communities shown in relation to the *x* ('Dry–Wet') and *y* ('Warm–Cool') axes. (a) Average acidity of the A layer of the soil in pH units. (b) Average water-retaining capacity of the A layer of the soil as percentage of dry weight. The southern wet forest (SW) has been disregarded in drawing the contours (from Curtis, 1959, by courtesy of The University of Wisconsin Press)

191

an evident relationship between the vegetation and the environmental factor plotted. This procedure is a very flexible one, allowing for the ready test of correlation between vegetation and any factor for which data are available. It should perhaps be emphasized that, of vegetation and factor, no indication is obtained of which is cause and which effect; other criteria must be used to determine this.

The techniques considered so far are purely phytosociological; arrangement of stands, whether classification or ordination, is based entirely on their species composition. An alternative approach is to base the arrangement entirely or in part on environmental gradients; in effect it is site characteristics that are used in drawing up a classification or ordination and vegetation that is shown to form a corresponding pattern rather than the reverse. Such an approach has been widely used, to a greater or lesser extent, in varied contexts. Many foresters are accustomed to work mainly in terms of site characteristics (see, for example, various contributions in Hustich, 1960). Various workers have attempted to arrange classificatory units in a one or more dimensional framework of gradients of moisture regime, altitude etc. (see, for example, Gams, 1961). Examples of more detailed investigations are the work of Kittredge (1938) who arranged stands containing *Populus tremuloides* according to soil characteristics and examined the relationship of species occurrence to the soil classes established, and of King (1962), who arranged stands of *Festuca–Agrostis* grassland in an ordination according to soil features. Dagnelie (1960) has suggested the factor analysis of sites, using environmental features as 'tests' of underlying factors; species distribution may then be plotted on the resulting ordination. The most thorough examination of the potentialities of the approach is, however, that of Whittaker (1952, 1954, 1956, 1960).

Whittaker's methods were elaborated in relation to two mountainous areas bearing mainly forest vegetation. Initial ordination was in relation to a 'moisture gradient' defined in terms of a physiographic series from deep ravines with flowing streams to open south facing slopes, i.e. from the most mesic to the most xeric sites. As Whittaker has pointed out, this 'moisture gradient' is not a gradient of moisture alone but includes a complex of correlated environmental gradients. On the results of placing stands in relation to this gradient, species were selected which had a clear mode in representation either towards one or other end of the gradient or around the middle. These species were used to place stands more precisely by a technique of weighted averages similar to that used independently by Curtis and McIntosh (1951) and Ellenberg (1950–54). For a particular stand the number of stems was multiplied by 0 for mesic, 1 for submesic, 2 for subxeric and 3 for xeric species and the total divided by the number of stems. The resulting range of compositional index was divided into a number of steps along the gradient. In his later study (1960) Whittaker also tested a technique of placing

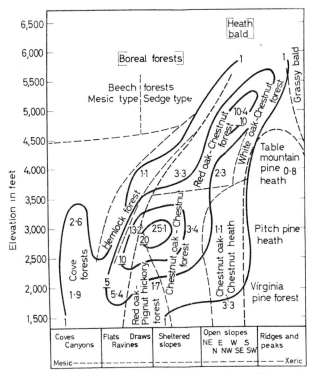

FIGURE 35. An example of gradient analysis. Stems of *Hamamelis virginiana* as percentage of total stems in stands in the Great Smoky Mountains in relation to gradients of elevation and moisture conditions. Figures are values for percentage of stems. Heavy lines are 'isorithms' (lines of equal percentage contribution to stands). Broken lines indicate the approximate boundaries of vegetation types in relation to the gradients studied (from Whittaker, 1956, by courtesy of *Ecol. Monogr.*)

by calculation of a coefficient of similarity with a stand at one extreme of the gradient, but concluded that this was less satisfactory than the weighted averages technique. Although based on different initial assumptions Whittaker's technique is closely similar to that of Curtis and McIntosh (1951), the difference being in the method of obtaining the weighting applied to individual species. For the vegetation examined it seems likely that the method of Curtis and McIntosh would give similar results to those obtained by Whittaker.

In his study of the Great Smoky Mountains Whittaker (1956) examined successive altitudinal belts in terms of the moisture gradient and thus produced a two-dimensional ordination with one axis representing the moisture gradient and the other elevational gradient. Representation of individual species could then be plotted on the ordination (FIGURE 35.) In his later study of the Siskiyou Mountains Whittaker (1960) found it necessary to use separate

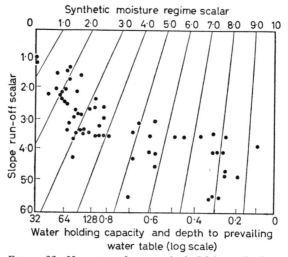

FIGURE 36. Nomogram for a synthetic Moisture Regime scalar (see text). Dots indicate the positions of samples in Loucks' (1962) study of forest communities in New Brunswick (from Loucks, 1962, by courtesy of *Ecol. Monogr.*)

two-dimensional ordinations for the vegetation on the main rock types present, giving, in effect, a three-dimensional arrangement. He concluded, in accord with Curtis and his associates, that variation in the vegetational composition of stands is essentially continuous.

Loucks (1962), in a study of forest communities in an area of varying topography in New Brunswick, has considered the construction of environmental gradients more critically. He has used scalars to integrate various environmental measures. A scalar is a numerical index relating two or more measures in such a way that the same value of the index reflects analogous relationships to, in this case, vegetation*. Their construction may be illustrated from one stage in Loucks' derivation of a moisture regime scalar (FIGURE 36). The horizontal axis in the figure is based on depth of water table where the latter is less than 7 ft. below the surface. Where the water table is deeper the moisture status is expressed in terms of water holding capacity of the soil. The vertical axis is a previously constructed scalar representing the proportion of heavy rain lost by run-off; this is derived from consideration of angle of slope and position on slope, which affects the amount of water received by 'run-off' from higher up the slope. The boundaries between segments of the moisture regime scalar are such that where the water table is high the value is little affected by the run-off scalar, but where the water table is low, the value is strongly dependent on the run-off scalar.

Loucks summarizes the environment in terms of the three scalars —moisture regime, nutrient status and local climate. He is then able to place stands, or, more accurately the sites occupied by the stands, in a three-dimensional ordination, in which species distribution may be plotted. He has also constructed a phytosociological ordination of the stands by the method of Bray and Curtis and has found a reasonably satisfactory agreement between the two ordinations in the relative placing of stands. He has pointed out that if the environmental variation can be adequately accounted for by position within a cube derived from three environmental gradients, the stands which resemble each other least are likely to be found at opposite ends of a diagonal of the cube, so that the first axis of a phytosociological ordination corresponds broadly to such a diagonal. The second and third axes, at right angles to the first, will similarly tend to lie along diagonals of the environmental ordination. Thus, if an environmental factor is plotted on a phytosociological ordination, the axis of variation may be expected to run diagonally; this is borne out by such data as are available (see, for example, Curtis, 1959).

It thus appears that, at least in some cases, ordination by environmental gradients and ordination by phytosociological inter-stand distance give comparable results. It is appropriate to consider whether one approach is to be preferred. In theory, phytosociological ordination would seem to be a more flexible approach; once the ordination is derived any environmental factor for which data are available can be plotted on to it and the resulting pattern examined. Environmental ordination involves a preliminary selection of the two or three gradients which are believed to be

* On the theory and use of scalars, see Torgerson (1958).

most important in determining the vegetation. An error of judgment at this stage may result in a much less efficient ordination, and it is significant that the successful environmental ordinations of Whittaker and Loucks have all been in regions where the main controlling environmental gradients are pronounced and obvious. Maycock and Curtis (1960) have examined data from the conifer-hardwood forest of the Great Lakes region both by the use of moisture gradients and by a three-axis phytosociological ordination and concluded that 'each treatment supported the results of the other'. Bray (1961) has derived various linear ordinations from the same, rather limited, set of data for four transects radiating from an isolated pair of trees in Wisconsin prairie. These included two vegetational ordinations, by the technique of Bray and Curtis and a modification of that of Curtis and McIntosh, two 'locational' ordinations, based on elevation and on linear distance from the trees, three environmental ordinations, based on soil density, on water retaining capacity of the soil and on light intensity, and, as a control, a random arrangement of stands. Each gradient was divided into five equal steps and the mean frequency of each species calculated for each step. A test of relative informativeness was based on summation for all species of the difference between 'order' (the difference between the maximum value for a species observed in any step of the gradient and the minimum value or values on either side of it) and 'disorder' (the amount by which species values must be changed to show a consistent trend from maximum to minimum value or values). On this test the two vegetational ordinations were the most informative and were fairly closely followed by those based on elevation and light, with the remainder proving much less informative. Various other tests agreed in showing the same four ordinations in the leading positions, though they varied in their exact placing. In view of the limited range of data, and the use of only linear ordinations the results are not conclusive; there is, moreover, room for difference of opinion which is the best test of informativeness. Bray concludes that 'the interrelatedness of causal factors and the consequent inadequacy of single factor explanations so often noted in ecologic writing suggest that there is a low probability of finding a gradient which is more complex' (i.e. informative) 'than one based initially on vegetational relationships'. This seems a valid conclusion, but further comparisons of different ordinations of one set of data are needed before it is accepted as a guiding principle.

Recently Whittaker (1960) and Beschel and Webber (1962) have considered the relative continuity of vegetation at different positions along environmental gradients, using similarity or correlation coefficients between adjacent stands. In view of the difficulty of disentangling point-to-point vegetational differences along an environmental gradient from varying steepness of the gradient itself, it seems unlikely that such investigations can contribute greatly to assessing the reality or otherwise of vegetational classes. When applied to a gradient observed on the ground (Beschel and

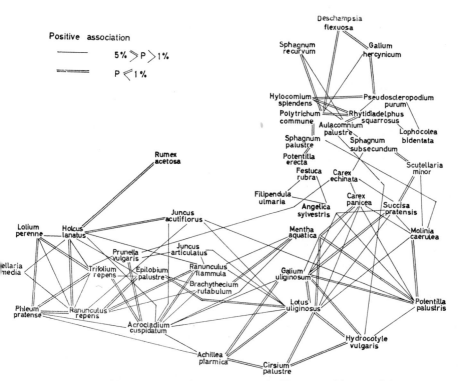

FIGURE 37. Positive associations between species in 99 communities containing *Juncus effusus* in North Wales (from Agnew, 1957, unpublished)

Webber), they may well give a useful indication of positions along the gradient which will repay further, more detailed, investigation.

An approach to communities somewhat different from those so far considered centres on the identification of groups of species showing mutual positive association. The members of such a group are presumed to have similar ecological characteristics or, at least, similar characteristics in relation to the controlling factors within a

197

particular set of stands. The ultimate objective of this approach is to use information from other sources on the ecological behaviour of some species to evaluate both the ecological characteristics of other species and the controlling factors of major importance in particular stands.

The preparation of a matrix of correlation coefficients is a necessary preliminary to R-techniques of factor analysis and information on the grouping of species is produced in the course of the analysis (FIGURE 27). Various workers have made more direct, though less precise, use of association between species. Some have extracted groups of associated species, e.g. Tuomikoski (1942) on the basis of correlation coefficients between frequencies of different species and Vasilevich (1961) on presence and absence. Others have found it useful to prepare diagrams, such as that shown in FIGURE 37, representing all the positive associations found, e.g. Agnew (1957, 1961), Hopkins (1957), Welch (1960) and McIntosh (1962), all of whom used presence and absence of species. De Vries (1953, 1954a), using $\sin\left(\dfrac{\chi^2}{\mathcal{N}} \times 90°\right)$ as a measure of strength of association has attempted to make the distance between species proportional to the strength of the associations between them. Not very surprisingly, this proved impossible in a two-dimensional frame and he was forced to include species in more than one position.

The use made of such ecological groupings of species has been varied. As has already been pointed out (p. 164), in some cases they have been made the basis of an indirect quantitative ordination or classification of stands. Welch (1960) has used an association diagram as the basis of a subjective delineation of associes. De Vries (1953, 1954a) has drawn conclusions on the species combinations likely to be most successful in grasslands. The utility of association diagrams is that they present a large amount of information in a form in which it is readily appreciated. Their limitation is that they only present the information; the element of simplification and generalization inherent in classificatory and ordinational techniques is lacking.

Fager (1957) has proposed basing groups on the degree to which species 'form a nearly constant part of each other's biological environment', and has used as an index of affinity the ratio of the observed number of joint occurrences of species to the expected value $\mathcal{N}_A\mathcal{N}_B/(\mathcal{N}_A + \mathcal{N}_B)$. He presents a table of 5 per cent points for selected values of $\mathcal{N}_A$, $(\mathcal{N}_A + \mathcal{N}_B)$, and $\mathcal{N}_B/\mathcal{N}_A$. This approach was developed in relation to animal communities, where direct interspecific relationships are likely to be more important. It is scarcely applicable to communities of higher plants but might conceivably be valuable in investigations of, for example, the micro-flora of soils.

Williams and Lambert (1961a) have suggested a more precise method of classification of species into ecological groups. This takes the form of an 'inverse' association-analysis. The procedure is identical with that of the 'normal' association-analysis already described but the data used are the values of $\sqrt{\chi^2/N}$ for comparison of all possible pairs of stands in terms of the number of species present in one, both, or neither stands. Division is then made on the stand showing the maximum total $\sqrt{\chi^2/N}$. Williams and Lambert point out a possible limitation to be borne in mind in interpreting the resulting groups. By the nature of the technique used the variates whose association is being examined are, in effect, standardized, i.e. brought to a common mean of zero and a common standard deviation of unity. Thus, in normal analysis differences of 'abundance' between species (number of stands occupied) are eliminated but differences of 'richness' between stands (number of species per stand) remain and may affect the analysis. Conversely, in inverse analysis, differences of richness are eliminated but differences of abundance remain. In inverse analysis groups are likely to reflect, in part, relative abundance of species; this is not, however, necessarily misleading as the most abundant species in this sense are likely to be those with the widest ranges of tolerance in respect of the environmental range of the stands analysed. Similarly differences in richness in normal analysis are likely to reflect real differences in habitat conditions of stands.

Williams and Lambert have, finally, attempted to bring together the results of normal and inverse association-analysis by assessing how far a species grouping is linked with a particular habitat grouping of stands (Williams and Lambert, 1961b; Lambert and Williams, 1962). The data are subjected to both direct (R) and inverse (Q) association-analysis. The species-site records are then arranged in a two-way table (FIGURE 38) in which the R-groups are delimited by the vertical boundaries between cells and the Q-groups by the horizontal boundaries. The division parameters in the two divisions are shown in the figure by open circles. (There is an ambiguity in the extraction of R-group 3 between species 15, 18 and 23, any one of which could have been used with the same result.) Each R-group is subjected to Q-analysis, and the site on which division would be made is marked by a marginal arrow at the top of the table; similarly, the species on which Q-groups would be divided by R-analysis are marked at the left of the table. The

	Species	Site 3	8	13	19	20	7	10	11	17	2	9	12	14	1	4	5	6	15	16	18
	(group)	(1)					(2)				(3)				(4)						
A	1 Calluna vulgaris	+	+	+	+	·	+	+	+	·	+	+	+	+	+	+	·	o	+	+	+
A	4 Erica cinerea	+	+	+	·	·	+	·	+	+	·	·	·	·	+	·	·	o	·	+	+
A	3 E.tetralix →	·	+	·	+	·	·	·	+	+	+	+	+	+	+	+	+	o	+	+	+
A	2 Molinia caerulea	+	+	+	+	+	+	+	+	+	+	+	+	+	+	+	+	o	+	+	+
A	10 Polygala serpyllifolia	+	·	·	+	·	·	·	+	+	·	·	·	·	·	+	·	o	+	+	+
A	8 Ulex minor	+	·	·	+	+	·	·	·	+	·	·	·	·	·	·	·	o	+	+	+
B	23 Drosera intermedia	·	·	·	·	·	·	·	·	·	o	o	o	o	·	·	·	·	·	·	·
B	18 D.rotundifolia	·	·	·	·	·	·	·	·	·	o	o	o	o	·	·	·	·	·	·	·
B	15 Eriophorum angustifolium	·	·	·	·	·	·	·	·	·	o	o	o	o	·	·	·	·	·	·	·
B	27 Juncus squarrosus	·	·	+	·	·	·	·	·	·	+	·	o	·	·	·	+	·	·	·	·
B	32 Pinus sylvestris	·	·	·	·	·	+	·	·	+	·	·	o	·	·	·	+	·	·	+	·
B	6 Trichophorum caespitosum →	·	·	·	·	·	·	·	·	·	+	+	o	+	+	+	·	·	+	+	·
C	9 Agrostis setacea	+	·	·	+	+	o	·	·	+	·	·	·	·	·	·	·	·	·	·	+
C	36 Carex pilulifera	+	+	+	+	·	o	+	+	+	·	·	·	·	·	·	·	+	·	·	+
C	21 Festuca ovina	·	+	+	·	·	o	·	+	·	·	·	·	·	·	·	·	·	·	·	·
C	11 Potentilla erecta →	+	+	+	+	+	o	·	+	+	·	·	·	·	·	·	·	·	·	·	·
C	5 Pteridium aquilinum	+	+	+	·	·	o	+	+	·	·	·	·	·	·	·	·	·	·	·	·
C	43 Quercus robur	·	·	·	·	·	o	o	o	o	·	·	·	·	·	·	·	·	·	·	·
D	14 Carex panicea →	+	·	·	+	+	·	·	·	·	·	o	·	+	·	·	·	·	·	·	·
D	19 Juncus acutiflorus	·	·	·	·	·	·	·	·	·	·	o	·	+	·	·	·	·	·	·	·
D	17 Narthecium ossifragum	·	·	·	·	·	·	·	·	·	·	o	·	+	·	·	·	·	·	·	·
D	13 Pedicularis sylvatica	·	+	·	+	·	·	·	·	·	·	+	o	·	·	·	·	·	·	·	·
E	56 Anemone nemorosa	·	o	·	·	·	·	·	·	·	·	·	·	·	·	·	·	·	·	·	·
E	58 Anthoxanthum odoratum	·	o	·	·	·	·	·	·	·	·	·	·	·	·	·	·	·	·	·	·
E	44 Betula verrucosa	·	o	+	·	·	·	·	+	·	·	·	·	·	·	·	·	·	·	·	·
E	57 Campanula rotundifolia	·	o	·	·	·	·	·	·	·	·	·	·	·	·	·	·	·	·	·	·
E	12 Carex binervis	·	o	·	·	+	·	·	·	·	·	·	·	·	·	·	·	·	·	·	·
E	61 Castanea sativa	·	o	·	·	·	·	·	·	·	·	·	·	·	·	·	·	·	·	·	·
E	46 Cerastium vulgatum	·	o	+	+	·	·	·	·	·	·	·	·	·	·	·	·	·	·	·	+
E	35 Galium hercynicum	+	o	·	+	·	·	·	+	·	·	·	·	·	·	·	·	·	·	·	·
E	37 Hieracium pilosella →	o	o	o	o	o	·	·	·	·	·	·	·	·	·	·	·	·	·	·	·
E	41 Hypericum humifusum	+	o	·	·	·	·	·	·	·	·	·	·	·	·	·	·	·	·	·	·
E	38 Hypochaeris radicata	·	o	+	+	·	·	·	·	·	·	·	·	·	·	·	·	·	·	·	·
E	50 Lathyrus montanus	·	o	·	·	·	·	·	·	·	·	·	·	·	·	·	·	·	·	·	·
E	48 Lonicera periclymenum	·	o	·	·	·	·	·	+	·	·	·	·	·	·	·	·	·	·	·	·
E	54 Lotus corniculatus	·	o	·	+	·	·	·	·	·	·	·	·	·	·	·	·	·	·	·	·
E	55 L.uliginosus	·	o	·	·	·	·	·	·	·	·	·	·	·	·	·	·	·	·	·	·
E	52 Luzula multiflora	·	o	+	+	·	·	·	·	·	·	·	·	·	·	·	·	·	·	·	·
E	45 Orchis? ericetorum	·	o	·	·	·	·	·	·	·	·	·	·	·	·	·	·	·	·	·	·
E	39 Sieglingia decumbens	+	o	·	+	+	·	·	·	·	·	·	·	·	·	·	·	·	·	·	·
E	34 Succisa pratensis	+	o	·	·	·	·	·	·	·	·	·	·	·	·	·	·	·	·	·	·
E	49 Teucrium scorodonia	·	o	·	·	·	·	·	·	·	·	·	·	·	·	·	·	·	·	·	·
E	53 Veronica chamaedrys	·	o	·	+	·	·	·	·	·	·	·	·	·	·	·	·	·	·	·	·
F	+ 33 other species																				

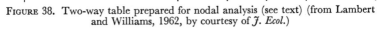

FIGURE 38. Two-way table prepared for nodal analysis (see text) (from Lambert and Williams, 1962, by courtesy of *J. Ecol.*)

Nodal analysis table:

Species	① 8 19 3 13 20	② 7 11 17 10	③ 2 12 9 14	④ 1 4 5 6 15 16 18
3 ERICA TETRALIX	●● · · ·	· ●● ·	●●●●	●●●●●●
1 Calluna vulgaris	+ + ● + +	+ + ● +	●+ + +	+●+ + + + +
2 Molinia caerulea	+ + ● + +	+ + ● +	●+ + +	●●+ + + + +
10 Polygala serpyllifolia	+ + ● · ·	· + · ·	· · · ·	· ●+ + + +
4 Erica cinerea	+ + ● + ·	+ + ● ·	· · · ·	+ · · + + +
3 Ulex minor	· + · · +	· · · ·	· · · ·	· · + + + ·
23 Drosera intermedia	· · · · ·	· · · ·	●+ + +	· · · · · · ·
18 D. rotundifolia	· · · · ·	· · · ·	●+ + +	· · · · · · ·
15 Eriophorum angustifolium	· · · · ·	· · · ·	●+ + +	· · · · · · ·
27 Juncus squarrosus	· · · + ·	· · · ·	●+ · ·	· + · · · · ·
6 TRICHOPHORUM CAESPITOSUM	· · · · ·	· · · ·	●●●●	· + + · · + ·
32 Pinus sylvestris	· · · · ·	+ · + ·	· + · ·	· · + · · + ·
5 Pteridium aquilinum	+ · ● · +	+ + · +	· · · ·	· · · · · · ·
9 Agrostis setacea	· + ● · +	+ + · +	· ~ · ·	· · · · · · +
36 Carex pilulifera	+ + ● + ·	+ + · +	· · + ·	· · · · · · +
11 POTENTILLA ERECTA	●●●●●	●+ + ·	· · · ·	· · · · · · ·
43 Quercus robur	· · · · ·	+ + ●+	· · · ·	· · · · · · ·
21 Festuca ovina	+ · · + ·	+ + · ·	· · · ·	· · · · · · ·
14 CAREX PANICEA	· + + · +	· · · ·	●●· ·	· · · · · · ·
19 Juncus acutiflorus	· · · · ·	· · · ·	○+ · ·	· · · · · · ·
17 Narthecium ossifragum	· · · · ·	· · · ·	○+ · ·	· · · · · · ·
13 Pedicularis sylvatica	+ + · · ·	· · + ·	○· · ·	· · · · · · ·
56 Anemone nemorosa	+ · · · ·	· · · ·	· · · ·	· · · · · · ·
58 Anthoxanthum odoratum	+ · · · ·	· · · ·	· · · ·	· · · · · · ·
44 Betula verrucosa	+ · · + ·	+ · · ·	· · · ·	· · · · · · ·
57 Campanula rotundifolia	+ · · · ·	· · · ·	· · · ·	· · · · · · ·
12 Carex binervis	+ · · · +	· · · ·	· · · ·	· · · · · · ·
61 Castanea sativa	+ · · · ·	· · · ·	· · · ·	· · · · · · ·
46 Cerastium vulgatum	+ + · + ·	· · · ·	· · · ·	· · · · · · +
35 Galium hercynicum	+ + ● · ·	· + · ·	· · · ·	· · · · · · ·
57 HIERACIUM PILOSELLA	●●●●●	●· · ·	· · · ·	· · · · · · ·
41 Hypericum humifusum	+ · ● · ·	· · · ·	· · · ·	· · · · · · ·
39 Sieglingia decumbens	+ + ● · +	· · · ·	· · · ·	· · · · · · ·
34 Succisa pratensis	+ · ● · ·	· · · ·	· · · ·	· · · · · · ·
38 Hypochaeris radicata	+ + · · +	· · · ·	· · · ·	· · · · · · ·
50 Lathyrus montanus	+ · · · ·	· · · ·	· · · ·	· · · · · · ·
48 Lonicera periclymenum	+ · · · ·	+ · + ·	· · · ·	· · · · · · ·
54 Lotus corniculatus	+ + · · ·	· · · ·	· · · ·	· · · · · · ·
55 L. uliginosus	+ · · · ·	· · · ·	· · · ·	· · · · · · ·
52 Luzula multiflora	+ + · + ·	· · · ·	· · · ·	· · · · · · ·
45 Orchis ?ericetorum	+ · · · ·	· · · ·	· · · ·	· · · · · · ·
49 Teucrium scorodonia	+ · · · ·	· · · ·	· · · ·	· · · · · · ·
53 Veronica chamaedrys	+ + · · ·	· · · ·	· · · ·	· · · · · · ·
Ⓕ 33 other species	+			

FIGURE 39. Nodal analysis of data shown in FIGURE 38 (see text) (from Lambert and Williams, 1962, by courtesy of *J. Ecol.*)

aim of the subsequent operations, which Lambert and Williams have termed 'nodal analysis', is to delineate groups of species-site records which would appear in either analysis, i. e. 'to explore the extent of coincidence between the two forms of the analysis, with a view to extracting the genuinely central species-in-habitat coincidences round which the population under examination may be considered to be varying'.

In an R-group the greatest likelihood of coincidence lies in those species which are represented in the site (i') on which the group would be divided by Q-analysis. Similarly in a Q-group the greatest likelihood of coincidence lies in those sites containing the species (a') on which the group would be divided by R-analysis. The sites i' and species a' thus represent 'coincidence parameters'. Any one cell of the two-way table may not contain any records pertaining either to the appropriate a' or to the appropriate i', e.g. 1B (FIGURE 39); if so it offers no evidence of coincidence. If any one of the sites appertaining to a cell contains all the species of the group covered by the cell, then a' must be represented by at least one record, though i' may not be. It is thus possible to limit the area of coincidence in the cell to those sites containing a', i.e. part of an R-group is defined giving an *R-subnodum* as in 3D. Conversely, if any one species is represented at all the sites of a cell, i' must be present though a' may not be and it is possible to limit the area of coincidence to those species represented in i' giving a *Q-subnodum*, e.g. 1C, 3A. If a' and i' are both present and a' is represented at i' then an area of coincidence limited in both directions, a *nodum*, is defined; e.g. 1E, 2A, 2C, 3B, 4A. In cell 1A, although it includes species represented at all sites and sites containing all species, the a', *Erica tetralix*, is not present at the i', site 3, and two intersecting subnoda, rather than a full nodum results. Clearly a full nodum is the most clearly defined and concentrated group. Williams and Lambert distinguish two levels of subnoda. In *major R-subnoda* a site containing all the species of the subnoda is either i' (not represented in FIGURE 39) or a division parameter of the Q-analysis (e.g. 3D). In *major Q-subnoda* a species represented at all sites of the subnodum is either a' (e.g. 1C, 3A), or a division parameter of the R-analysis (not represented in FIGURE 39). These represent a higher probability of coincidence than *minor subnoda* defined only by a' or i' (e.g. the intersecting subnoda of cell 1A).

Lambert and Williams have tested this nodal analysis on two sets of data, the rather simple set, part of the analysis of which is shown in FIGURES 38 and 39, and a more complex set from marsh communities in the Norfolk Broads. In both cases they were able to derive ecologically valuable information and the method appears to be of considerable potential value. As they point out, the concept of species-in-habitat units is not new; it appears in such familiar vegetational terms as 'reedswamp' and 'wet heath'. It has not, however, previously been placed on an objective basis.

Various techniques of classification and ordination have been outlined above. The choice of technique to be used in a particular case is not always easy. It will in part be dictated by the type of data that are available or can conveniently be collected, and by the computing facilities available. There is, however, another consideration involved. Although the techniques can all be considered as either classificatory or ordinational, they have been developed with very varying objectives in mind. Thus association-analysis, though resulting in a formal hierarchical classification, has been developed with the detailed ecological interpretation of a particular set of vegetational samples in mind, and the groupings that result are not necessarily representative of abstract vegetational units. The objective is thus very different from that of most formal systems of classification of vegetation and the results should not be compared with them. Again, factor analysis, in the strict sense, is most useful, as has been shown above, when applied to a very limited range of vegetation, and is unlikely to be justified in broader surveys. Thus, the choice of technique will also rest on the kind of information that is required. At the present stage it is not possible to make any firm general recommendations. It is, however, encouraging to note that in the relatively few cases where two or more techniques have been applied to the same data, they have yielded consistent results.

There remains one topic which may appropriately be considered in this chapter, the assignment of additional stands to their correct position in a previously established classification. It is always possible to reclassify the whole augmented set of stands but it may be more practical to accept the definitions of groups previously derived and place additional stands accordingly. If the groups overlap in their recorded characteristics, there is inevitably some possibility of misclassification of a stand having characteristics within the range of overlap. This risk also applies if a stand is found which is intermediate between two groups previously considered not to overlap. Nevertheless, it is possible to assign the stand to its group in such a way that the probability of misclassification is minimized.

The concept of *fidelity* derives from Braun-Blanquet (see Braun-

Blanquet, 1951), who put forward the idea that certain species are more or less faithful to particular communities. In practice, the concept of fidelity is useful only within circumscribed limits, either within a number of related communities or within a geographically defined region. As the concept is used by the Zürich–Montpellier school, the determination of the faithful species of a community and of the degree of their fidelity is a subjective process, and the various grades of fidelity proposed were originally defined in general terms only.* Some arbitrary requirement, to be satisfied before a species is accepted as showing fidelity to a particular community, may be adopted. This may consist of a minimum difference in percentage presence in stands between the community in question and the one in which the species attains the next highest figure, or a minimum difference in abundance of the species between the community to which it is supposedly faithful and that in which it has the next highest value, e.g. Curtis and Greene (1949), in a study of Wisconsin prairies, accepted as preferential species those with a presence value 10 per cent higher than in any other type. Braun-Blanquet (1951) himself later accepted a scheme defining grades of fidelity in terms jointly of percentage presence and cover value.

Goodall (1953b) has attempted to place determination of fidelity on a more objective basis. The appropriate procedure depends on whether the communities have been described in terms of presence or absence only or by quantitative measures. He points out that testing fidelity of a species to a particular community in the former case is essentially the same as testing the association between two species in a set of samples. If a species is believed to be faithful to community A, and community B is the one in which it occurs next most commonly, we have a contingency table

	Community		Total
	A	B	
Species present	a	b	$a + b$
Species absent	c	d	$c + d$
Total	$a + c$	$b + d$	$a + b + c + d = N$

which may be tested for departure from random expectation (see Chapter 4), and which gives an unbiased indication whether or not

---

* For a readily available account of the methodology of the Zürich–Montpellier school, reference may be made to Poore (1955-6).

the species does tend to occur more commonly in community A than in community B. Goodall suggests, as a suitable index of degree of fidelity, the difference in frequency of the species in the two communities expressed as a proportion of the lesser value, i.e.

$$\frac{\dfrac{a - \frac{1}{2}}{a + c} - \dfrac{b + \frac{1}{2}}{b + d}}{\dfrac{b + \frac{1}{2}}{b + d}} = \frac{(a - \frac{1}{2})(b + d)}{(b + \frac{1}{2})(a + c)} - 1.$$

This ranges from 0 to infinitely high values. Yates's correction for continuity (subtraction of $\frac{1}{2}$ from values below expectation and addition of $\frac{1}{2}$ to values above expectation) is made for the same reasons as in the calculation of χ^2. It may result in a value slightly below zero, owing to over-correction and, if so, the fidelity index should be taken as zero.

The calculation of the index may be illustrated from Adriani's data as quoted by Goodall. *Obione portulacoides* occurred in 20 out of 22 stands of Artemisietum maritimae and in 13 out of 24 stands of Puccinellietum maritimae, the community of the group studied in which it was next most frequent. The contingency table is

		Artemisietum	Puccinellietum	Total
Obione portulacoides	present	20	13	33
	absent	2	11	13
	Total	22	24	46

giving a value of $\chi^2 = \dfrac{46\,(19{\cdot}5 \times 10{\cdot}5 - 13{\cdot}5 \times 2{\cdot}5)^2}{33 \times 13 \times 22 \times 24} = 5{\cdot}938$

$$\chi = 2{\cdot}44$$

which corresponds to a probability of about 1 per cent (from Fisher and Yates (1943) Table VIII). The index of fidelity is

$$\frac{19{\cdot}5 \times 24}{13{\cdot}5 \times 22} - 1 = 0{\cdot}57.$$

This approach may be used equally well for data of 'presence' (occurrence in stands, which, though widely used, is not a satisfactory form of record as the chance of occurrence of a species is partly dependent on the size of the stand), constancy in the sense of the Zürich–Montpellier school (occurrence in uniform-sized samples from different stands), or frequency (totalled over all stands of a group). The resulting index of fidelity has only a limited value in assigning a further stand to its group. If such a stand has several species all with high fidelity value to one group, and no species with high fidelity value to any other group, its assignment is straightforward. If a stand has several species of moderate fidelity value to different groups, its classification is much more difficult. Goodall has suggested using an index of *indicator*

TABLE 24

INDICATOR VALUE OF FIVE SPECIES FOR FOUR GROUPINGS IN MALLEE (from GOODALL, 1953b, by courtesy of *Aust. J. Bot.*)

Grouping	Indicator value of species					Total indicator value for quadrat containing all five species
	Dodonaea bursariifolia	*Eucalyptus dumosa*	*Eucalyptus oleosa*	*Stipa variabilis*	*Vittadinia triloba*	
Cassia eremophila	—22·32	—3·77	+0·78	+0·43	—0·23	—25·11
Bassia uniflora	—2·17	—1·05	+0·80	+0·47	+0·21	—1·76
Dodonaea bursariifolia	+1·25	+0·80	—1·75	—107·73	—2·45	—109·88
Vittadinia triloba	+0·46	+0·34	—0·22	+0·76	+0·76	+2·10

value, comparable to the index of fidelity but depending on comparison of the frequency of occurrence of a species in a group to which it is supposedly faithful, with its frequency in all other groups. Thus, in his data for Mallee, mentioned above, *Eucalyptus dumosa* is present in two quadrats only of the 17 placed in the *Cassia eremophila* grouping but in 168 out of the 239 quadrats in other groupings, so that $a = 168$, $b = 2$, $c = 71$, $d = 15$, and the index of indicator value is $\dfrac{(168 - \frac{1}{2})\,(2 + 15)}{(2 + \frac{1}{2})\,(168 + 71)} - 1 = 3\cdot77$. The significance of the value is again tested from the contingency table. In this case *E. dumosa* is less frequent in the *Cassia eremophila* grouping than in the others taken together and the indicator value is read as −3·77, i.e. presence of *E. dumosa* in a quadrat is evidence against its belonging to the *Cassia eremophila* grouping. The use of indicator value rather than fidelity not only makes use of such negative information; it also takes into account cases in which a species is frequent in two or more groupings but occurs rarely in all others. In such a case the species would have low fidelity for all groupings, but a relatively high indicator value for the groupings in which it is frequent.

The use of indicator value may be illustrated from an example given by Goodall, again from his Mallee data. The five species, *Dodenaea bursariifolia, Eucalyptus dumosa, E. eleosa, Stipa variabilis* and *Vittadinia triloba,* have the indicator values for his four groupings shown in TABLE 24. The total indicator value for the five species is greatest for the *Vittadinia* groupings, and hence a quadrat containing all five species and no others would be assigned to this grouping.

Calculation of the total indicator value of a sample for various possible groupings is readily made, even if a large number of species is involved in their definition. If the groupings are defined in terms of frequency of a small number of species only, a more exact treatment may be used, which has the added advantage of furnishing an estimate of the probable degree of misclassification. This is due to F. E. Binet (see Goodall, 1953b) and takes the form of determining for each combination of species the grouping in which it has the greatest chance of occurrence. Suppose two communities are defined in terms of frequency in the following way:

	Community A (per cent)	Community B (per cent)
Species X	80	10
Species Y	70	20
Species Z	10	80

The combination XYZ has a probability of occurrence in community A, in a quadrat of the size used in determining frequency, of $0·8 \times 0·7 \times 0·1 = 0·056$. In community B the corresponding probability is $0·1 \times 0·2 \times 0·8 = 0·016$. A quadrat containing all these species will, therefore, be assigned to community A. Similarly, a quadrat containing Y and Z but not X will have a probability of occurrence of $0·2 \times 0·7 \times 0·1 = 0·014$ in community A and of $0·9 \times 0·2 \times 0·8 = 0·144$ in community B and will be assigned to the latter. The probabilities for the eight possible combinations of species, with the community to which each is assigned, are:

Species combination	Community A	Community B	Community to which assigned
XYZ	0·056	0·016	A
XY–	0·504	0·004	A
X–Z	0·024	0·064	B
–YZ	0·014	0·144	B
X—	0·216	0·016	A
–Y–	0·126	0·036	A
—Z	0·006	0·576	B
—	0·054	0·144	B

It will be seen that the combinations X–Z, –YZ, —Z, and ——, if found in quadrats rightly belonging to community A, will be assigned to community B. The probability of misclassification of quadrats drawn from community A is, therefore, 0·024 + 0·014 + 0·006 + 0·054 = 0·098 or 9·8 per cent. Similarly the probability of misclassification of quadrats drawn from community B is 0·016 + 0·004 + 0·016 + 0·036 = 0·072 or 7·2 per cent, giving a mean chance of misclassification of 8·5 per cent. This approach can be used for any number of groupings defined in terms of any number of species but rapidly becomes more cumbersome as the number either of groupings or of species is increased.

The groupings and the unknown stand may have been described in terms of cover, density, or other absolute measure, in which case Fisher's (1936) *discriminant function* is available. It is likewise available if a stand is being assigned to one of a number of groupings defined in terms of frequency. A discriminant function between two groupings is of the form $ax_1 + bx_2 + cx_3$ etc., where x_1, x_2, x_3, etc., represent values of the measurements used in defining the groupings (in this case the quantities of different species) and a, b, c, etc., are calculated to maximize the difference between the mean values of the function for the two groupings.[*] Mather (1946) gives a convenient account of the calculation of discriminant functions. Once the discriminant function has been determined for a pair of groupings its numerical value for any further sample may be calculated. The sample is assigned to that grouping whose mean value for the discriminant function it approaches most nearly. The degree of misclassification resulting from use of the discriminant function as the sole criterion can be determined, as can the increased degree of misclassification that results from a species being eliminated from the discriminant function. This gives an indirect indication of

[*] The discriminant function is related to the statistic D^2 (p. 123). Measurement of the difference between groupings in terms of a discriminant function is equivalent to the use of D^2.

the relative importance of the different species in definition of the groupings. If more than two groupings are involved it is normally necessary to calculate separate discriminant functions for each pair: but it is sometimes possible, without appreciable loss of precision, to calculate one function discriminating between, say, groupings A and B on the one hand and C on the other, and one further function discriminating between A and B, or even to calculate a single function discriminating between all three groupings.

CHAPTER 8

THE QUANTITATIVE APPROACH TO PLANT ECOLOGY

At the present stage of development of plant ecology, when the great value of a quantitative approach is becoming apparent, there is a tendency either to apply quantitative methods too enthusiastically, or to apply them to problems to which they are not, or not yet, appropriate. This is an example of a very general tendency when advances in technique are made, and one which may bring such advances into temporary disrepute. It is essential to keep a sense of perspective and to consider whether the time consumed by the use of quantitative instead of qualitative methods will give a commensurate increase in the value of the results obtained. Similarly, it may be a waste of time to apply a statistical test of significance to the difference between two means so widely different, and based on ranges of values so widely different, that the probability of their being drawn from the same population is negligible. It is sometimes necessary to avoid too sensitive a method for quite other reasons. This may be illustrated from studies of long-term change, e.g. the changing composition of grassland under a changed régime of grazing. With such vegetation the obvious measure to use is that of cover. At certain seasons of the year, particularly when active growth is starting in the spring, the cover of a species changes very rapidly; even day-to-day changes may be detectable. Moreover, if, as is commonly the case, neither the times of resumption of growth nor the periods of most active growth of different species are synchronized, even the relative cover of different species will change. Thus cover estimates will vary widely during the year as a result of seasonal change, quite apart from any long-term trend. Unless the greatest care is taken to ensure that yearly estimates of cover are made at comparable stages of seasonal development, the fluctuations of cover observed may be so great that any long-term trend is completely obscured unless a very long series of observations is made. If the vegetation remains unchanging for some period of the year, it may be possible to ensure sampling at comparable stages each year but the stage must be decided by

210

reference to the vegetation and not by the calendar date. If sampling at comparable stages is not possible, it is better to use a less sensitive measure, e.g. rooted frequency, which, while less sensitive to change in composition of the vegetation, is relatively unaffected by seasonal changes.

There are two other dangers to which ecologists using a quantitative approach are exposed, (1) such excessive preoccupation with technique that the ends in view are pushed into the background, and, related to this, (2) attempts to force ecological data into formal mathematical moulds. Preoccupation with technique is perhaps inevitable in a developing subject, but such work as that of Watt (1940–56) on the community structure in *Pteridium aquilinum*, Rutter (1955) on wet-heath vegetation and Curtis and his associates (see Chapter 7) on ordination, shows that it is possible to keep a firm grasp of the ecological problems, while making full use of quantitative techniques. The dangers of formalism, for which there is less excuse, have already been discussed in relation to non-random distributions and the attempts to fit mathematical series to them.

Just as a stage of qualitative investigation must precede a quantitative approach in a branch of science as a whole, so in examining a particular problem it is essential to grasp the broad aspects before starting quantitative observations. The selection of the site or type of vegetation to be examined in detail (in contrast to the positioning of samples within it) must be made subjectively, and the success or failure of the investigation may be decided by this selection. On the one hand the areas chosen for intensive study must be those most relevant to the problem. On the other hand any bias towards a preconceived interpretation must be avoided. Studies of community delimitation are particularly subject to such initial bias. It is clear that a delicate compromise must be effected in the choice of sites as between inclusion of such a wide range of sites that much of the resulting data is irrelevant, and concentration on sites that appear to support an initial hypothesis. Time and thought spent at this stage are rarely wasted. Unfortunately, some published work gives the impression that the sites examined have been selected almost arbitrarily. It need hardly be said that quantitative observations should not be made without a definite objective, though this may be only to answer a very simple and preliminary question such as 'Is there any difference in the density of a certain species between areas A and B?' The tendency to regard the collection of numerical data as an end in itself cannot be too much deplored.

Survey of the literature suggests that sampling is the least well understood aspect of quantitative ecology. Not infrequently observations have been wisely planned and the data skilfully analysed but the results are invalidated by the unsatisfactory sampling procedure used. Even in recent papers such phrases as 'quadrats were placed at equal intervals along the line to give random samples' are to be found. Such errors perhaps arise because most textbooks of biological statistics are written mainly with experimental work in mind, and in such work considerations of sampling are more straightforward so that the subject is treated rather perfunctorily. It behoves the ecologist to be especially careful in preparing a scheme of sampling. Some of the considerations involved, and discussed earlier, apply equally if it is only qualitative observations that are to be made. Thus the necessity of considering the frequency of different habitat conditions associated with individuals of a species in relation to their frequency in the area as a whole, discussed in relation to pH in Chapter 5, holds equally if the habitat is to be examined in a qualitative way only.

Many of the advances that may be expected as the result of using a quantitative approach, with judgements based on the methods of statistical analysis, will be apparent from the preceding chapters. They may be summed up as follows: (1) The quantitative approach allows the detection and appreciation of smaller differences. (2) By using suitable statistical tests the quantitative approach provides a sounder basis of judgement of the significance of differences observed. It thus becomes possible to find an answer to questions quite unanswerable by qualitative techniques, but this represents essentially an improvement on, and extension of, existing approaches to particular problems. It remains to be considered whether the quantitative approach can be expected to make any special contribution to ecological theory.

Ecology has developed as a branch of botany which is largely empirical, with few generalizations. Among the few integrating hypotheses those of succession and climax, and of the plant community as an organism, quasi-organism or superorganism have been especially prominent. These hypotheses, while they have proved fruitful, have remained essentially theoretical conceptions difficult of proof or disproof. Even the evidence for successional changes, commonly accepted as almost axiomatic, is largely circumstantial. The study of pattern, an aspect of vegetation dependent on quantitative methods for its detection in all but the most obvious cases, holds out promise, as will be shown below, of more definite assessment of these hypotheses.

There is an extensive literature concerned with the theory of climax, which has been reviewed by Whittaker (1953). Much of the controversy about the climax has been concerned with whether the climatic climax is the only true climax, of which the edaphic and biotic climaxes of some schools of thought are incomplete expressions, or whether more than one true climax can exist in the same climate. These arguments do not directly concern us here. Whatever view is taken of the proper limitation of the term, all climaxes have in common relative stability of composition and structure, in contrast to the changing composition and structure of seral communities. Stability can only be relative, for no vegetation is completely stable if viewed over a long enough period of time. We have seen that pattern may be present in a plant community on very varying scales. Pattern in part reflects the morphology and behaviour of the species concerned but it may also arise from control by environmental factors and then reflects the pattern of values of these influencing factors.

Consider different vegetation in two physically similar habitats A and B and suppose that in A environmentally determined pattern of vegetation is predominantly large scale, while in B this element of the pattern shows a concurrent smaller scale. With one proviso, to be considered below, the pattern of levels of the influencing factors is likely to be similar, since the habitats as a whole are physically similar. Assuming, as is reasonable in two similar habitats, that the same environmental factors are the principal influencing ones, the absence of smaller-scale pattern of vegetation in A can only mean that the constituent species of this vegetation are less sensitive to small differences in the influencing factors than those in B and that a greater difference in the level of influencing factors would be necessary to produce a detectable difference in composition. Or, put another way, the same difference in influencing factor in B is responsible for a greater difference in amount of species than in A, i.e. the intensity of the pattern is greater in B than in A.

If a small difference in level of the influencing factor is responsible for differences in composition of the vegetation, as in B, a relatively small general modification of that factor in the habitat as a whole, whether arising from secular change or from the reaction of the species present or entering, will produce marked change in the vegetation, some species becoming more abundant and others becoming less abundant or being eliminated. In A much larger changes in influencing factors will be necessary to produce noticeable change in the vegetation. The vegetation in A will, therefore, be more capable of surviving change in its environment than that in B,

213

i.e. it will be more stable. A criterion of stability, based on conditions at one time alone, is thus available. Within similar habitats the greater the average scale of environmentally determined pattern the more stable the vegetation. Further, if the scale of pattern is the same, the less intense it is the more stable is the vegetation. This approach to stability avoids the difficulty introduced by differing generation time of different species, inevitable when stability is judged solely on observation over a period of years. Definition of stability in terms of number of generations of individuals during which vegetation remains the same seems more reasonable than definition in terms of time alone. On this basis vegetation composed mainly of species with long generation time and which remains the same for one generation only, e.g. some forest types, is no more stable than that which changes after one generation of ephemerals. Change is, however, very difficult to detect by observation in the former case but immediately apparent in the latter.

The assumption of similar patterns of levels of environmental factors is subject to one qualification. The average intensity of pattern of environmental factors may be decreased and smaller scales perhaps eliminated by the reaction of the vegetation tending to equalize differences in environmental factors. Two obvious examples of such effects are the overlaying of mineral soil by the peat formed by vegetation, and the lessening of fluctuation in humidity and temperature brought about by the insulating effect of a vegetation cover. This equalizing of environmental differences will result in a greater spatial separation corresponding to a given difference in level of an environmental factor, so that a particular sensitivity of a species to an influencing factor will be reflected as a larger scale of pattern. The existence of this effect might appear in part to invalidate the thesis put forward above that larger scale of pattern reflects less sensitivity to environmental factors and hence greater stability. Further consideration shows that, while less sensitivity may not always be indicated, greater stability may still reasonably be assumed. The levelling out of differences in the environment by reaction implies that externally determined controlling conditions are largely replaced by vegetationally determined ones. Such vegetationally determined conditions are likely to prove to be inherently more stable than externally determined ones, for the effect of any external alteration is lessened by the pressure of reaction tending to maintain the *status quo*. Thus a change in climate tending to drier soil conditions will be less pronounced in an organic than in a mineral soil. The effects of reaction in increasing scale of pattern are likely to be more important in more complex communities and especially in those climaxes which are the endpoints of successions showing a pronounced change from extreme xerophytic or hydrophytic conditions to mesophytic ones.

The concept of succession is closely connected with that of climax. Similar conditions to those just discussed apply to individual species in the course of succession. When a species first invades an area of changing vegetation, unless for any reason dissemination of its propagules over the area has previously been impossible (e.g. by the absence of reproducing individuals in the neighbourhood), conditions must be just within its limits of tolerance. Otherwise it would have been able to invade the area at an earlier stage in succession. It will therefore be sensitive to environmental differences within the area and exhibit small-scale pattern. If changing conditions produced by reaction, partly of other species and partly of the invading species, tend to favour it, it will become less sensitive to the environmental differences present and exhibit correspondingly larger-scale pattern. If, on the other hand, changing conditions do not favour it, the scale of pattern will remain the same or even be reduced. Thus as long as conditions and, correspondingly, vegetational composition, continue to change, i.e. as long as succession continues, the vegetation, while containing some species for which conditions are at their optimum in the succession and therefore show maximum scale and minimum intensity of pattern, will also include invading species of small scale and high intensity of pattern, so that the average scale of pattern will be less than that in the corresponding climax, the stable endpoint of the succession.

The tendency to predominantly small scale of pattern in seral stages may be reinforced by reproductive or morphological features of the invading species. If colonization by the invading species is initially sparse, even within those parts of the area that permit establishment, the second generation individuals may be grouped around the initial colonizers, either because of their limited distance of seed dispersal or because spread is vegetative. These groups may even be smaller in average size than the favourable parts of the area, reducing the scale of pattern below that expected from the pattern of levels of the influencing factors. In succeeding generations the groups will increase in size as they spread to the boundaries of favourable parts of the area. As the groups come into contact with one another the smaller scale of pattern will disappear relatively suddenly. The remaining pattern will increase in scale as the areas favourable to the species widen (FIGURE 40). Even if conditions in the area as a whole rapidly become favourable to the species, a limit to the rate of spread, and hence of increase in scale of pattern, may be set by the reproductive behaviour of the species. If conditions continue to change in favour of the species, these areas may eventually coalesce, leading to complete disappearance of environmentally determined pattern. Alternatively the general

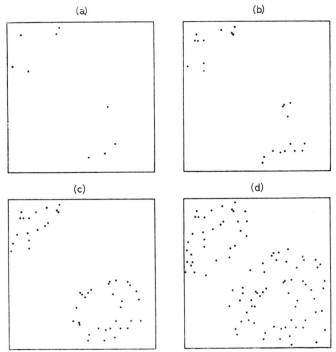

FIGURE 40. Change in pattern of a species in course of succession. (a) Initial colonization by scattered individuals in favourable parts of the area, giving pattern of corresponding scale. (b) Formation of small groups by reproduction from the initial colonizers, resulting in an additional smaller scale of pattern. (c) Disappearance of smaller scale of pattern as groups of individuals coalesce. (d) Increase in scale of remaining pattern as the proportion of the area favourable to the species increases consequent on reaction

trend of change may become again unfavourable so that the species begins to decline. In this case a maximum scale of pattern will coincide with maximum representation of the species in the vegetation. Since for most species the limits of tolerance for establishment are much narrower than for the survival of established individuals, the decline in representation of the species from the stage at which conditions are optimal is unlikely to be attended by any change in scale of pattern. At least, this will hold provided conditions become generally less favourable over the whole area, which is likely unless later invaders produce strongly localized reaction. Skellam (1952) has pointed out that if random elimination of individuals occurs in a randomly distributed population, the distribution remains random, and, more important, if individuals are eliminated randomly from a non-randomly distributed population, a random distribution will eventually be produced when the density is sufficiently reduced. Thus, although the scale of pattern will generally remain the same, its intensity will lessen as the species declines and eventually the species will attain a random distribution, apart from any morphological pattern. Whatever the exact course of events a species is likely to pass during succession through stages of successively greater average scales of pattern, culminating in randomness, but the stage at which randomness is reached may vary from that of maximum representation to the stage just before disappearance from the community.

This discussion of pattern in relation to climax and succession has been based on theoretical considerations. Supporting data is, as yet, scanty. Whitford (1949) demonstrated the lower intensity of pattern in the herbaceous species of climax woodland compared with successional stages. Greig-Smith (1952b) found increasing scales of pattern in successional stages in secondary rain forest. Kershaw (1958) found decreasing intensity and eventual elimination of environmentally determined pattern in *Agrostis tenuis* in reverting upland pastures as the sward approaches stability. Greig-Smith (1961d) found environmental pattern in *Ammophila arenaria* at the earlier but not the later stages of succession. If the thesis put forward is confirmed by further work, the examination of pattern would seem to offer much more objective criteria of stability than have hitherto been available. More definite information on the relative stability of communities would in turn facilitate investigation and understanding of the part played by gross environmental differences, particularly of climate, which is relatively little affected by reaction, in determining the broad features of vegetation.

The nature of the plant community has been the subject of much disagreement. Views have varied from that of an assemblage of individuals with similar tolerances having no specific effect on one another (Gleason, 1926) to that of a complex organism (Clements, 1916, etc.; Clements and Shelford, 1939; Phillips, 1934–5) in which the relationship between individuals is analogous to that between cells in an organism. Between these extremes a variety of

217

intermediate viewpoints has been adopted. Little attempt has been made to consider the implications of these widely differing hypotheses in terms of community structure. If the individualistic viewpoint is the correct one, then all point-to-point variation in the community must be due either to the controlling effect of variations in environmental factors or to chance, i.e. which species arrived at a point first. If the organismal viewpoint is correct the close relationships between individuals must have detectable effects on the patterns exhibited by the constituent species (Greig-Smith, 1952b). If there are numerous interactions between individuals, some at least of these interactions must affect the survival or otherwise of individuals concerned. If individuals of one species depress the survival chances of others of the same species establishing within a certain distance of them, the result, at least at high densities, would be a tendency to regular distribution. As we have seen, regular distributions are very rarely found in the field. If individuals increase the chance of survival of others of the same species near them, small-scale pattern will result. Proof that observed small-scale pattern is due to direct effect between individuals is difficult. It can only rest, in most cases, on the absence of any other evident cause. If individuals of one species increase or decrease the chances of survival of individuals of other species, the patterns of the influenced species or of both, if the effect is mutual, will likewise be affected. The effect will most readily be detected by association data from relatively small sample areas. Again, proof of the responsibility of direct effect between individuals can only rest on the absence of any other evident cause.

Data on pattern in relation to the nature of the community are scanty. What there are point to the individualistic concept as nearer the truth. At least there is no real evidence in support of the complex organism. Positive association at small scales is rare and can generally be explained in terms either of similar habitat requirements in areas where environmental pattern has a very small scale, or by common exclusion of the associated species by a third species. Negative association is much commoner but can generally be attributed to spatial exclusion (Chapter 4), i.e. the impossibility of two individuals growing in the same place, or to competition, i.e. (in this context) similar demands on the habitat by the two species (cf. Watt, 1955; Greig-Smith, 1961a; Kershaw, 1962a, b). These are, it is true, direct effects between individuals but scarcely of the highly specific kind implied by the concept of the complex organism. Against this present lack of evidence from pattern in favour of the complex organism may be set the production by some species of substances which are toxic to other species (Bonner, 1950). Though their existence is well established, it is still uncertain how important they are in field conditions. Muller (1953) has brought forward evidence that the effect of desert shrubs on annuals, pointed out by Went (1942) (see Chapter 5), arises more from the differential effect on the accumulation of organic matter in the soil than from their production of toxins. Evidence from ordination studies also points increasingly to the individualistic viewpoint of the community as being more satisfactory. (Cf. Chapter 7; see especially Curtis, 1959, and Whittaker, 1956, 1960.)

APPENDIX A

1. *Meteorological Data*

Correlation with meteorological data often presents the ecologist with difficult problems, especially in deciding which measure of the climatic factor is the appropriate one. Thus, temperature may be recorded as the annual mean, the mean daily minimum, the length of frost-free period in the year, the accumulated total time above a certain temperature in the year, or a variety of other possible measures, all of which are appropriate to particular investigations. The choice of measure to be used is generally determined by knowledge of the biology of the species concerned, and is beyond our present scope (cf. Salisbury, 1939). There are, however, points about the handling of the data, once the appropriate measure has been decided, which are worth considering.

Since different climatic factors are, within a limited region, often closely correlated with one another, the danger of inferring causal relationship from correlation is especially prominent. Further complications may result from the occurrence of trends of change in climate extending over a number of years. Thus, in agricultural data for yield, it is sometimes possible to show correlation between the yield of one year and the weather of the following year. These difficulties must be kept in mind in considerations of data for a period of years. Comparison of short-term changes, within one season, with changing weather is generally more straightforward. Such studies generally involve measures of performance, e.g. weight of foliage produced, number of seeds germinated, etc., by different dates. It is usually appropriate to consider the cumulative totals, both of the performance and of the meteorological measure, because the later plant performance may be affected not only by the immediately preceding weather but also by that earlier in the season. Analysis of such data may be considered by an example:

There was reason to believe that the germination of achenes of *Urtica dioica* was correlated with the amount of sunlight received after they had been moistened. TABLE 1 shows the total number of achenes germinated in three replicate lots of 100, at various intervals after moistening, together with the total hours of sunshine received. The regression of number of seeds germinated (y) on number of days from moistening (x_1) and hours of sunshine (x_2) is $y = 21.46 - 5.20x_1 + 2.74x_2$. The coefficient of x_1 is not significant but that of x_2 has a probability of less than 1%, indicating a real correlation between germination and hours of sunshine received. Indeed, as far as the evidence from this one experiment indicates, the amount of germination is independent of time since moistening. A significant proportion of the variance between dates is not accounted for by the regression, but this does not concern us in the present context.

TABLE 1

NUMBERS OF ACHENES OF *Urtica dioica* GERMINATED IN THREE REPLICATE LOTS OF 100

Days from moistening	Hours of sunshine to previous day	Total number of achenes germinated				Expected number of achenes germinated (from regression)
		I	II	III	Mean	
11	14.8	1	—	5	2.0	4.8
12	20.2	9	4	17	10.0	14.4
13	26.9	26	16	40	27.3	27.6
14	30.9	35	24	45	34.6	33.3
15	34.9	44	38	50	44.0	39.1
16	39.3	50	44	56	50.0	51.1
17	43.7	53	51	59	54.3	52.8
18	45.2	58	54	59	57.0	51.7
19	50.5	59	56	61	58.6	61.0
20	54.1	60	60	62	60.6	65.7
21	55.7	61	65	62	62.6	64.8

It is sometimes important to determine whether two microclimates can be regarded as the same or not. A useful technique of comparison was applied to several environmental factors, e.g. evaporation rates, by Harris *et al.* (1929) but appears to have been overlooked since. The evaporation rate was measured daily at four sites for a month. It varied widely from day to day at all the sites and graphs showing the changes at the different sites were very difficult to compare. In this context mean evaporation rate has very little relevance; the same mean could be given by widely differing régimes of evaporation. Harris *et al.* pointed out that if evaporation régimes at the two sites were identical, a plot of the evaporation rate at one site (E_1) against that at the other (E_2) on the same day should fit the equation $E_1 = E_2$. If the regression $E_1 = a + bE_2$ is calculated, and the value of the constant a is significantly different from 0, or that of b significantly different from 1, the null hypothesis of identity between the sites is disproved. This method of comparison provides a much more sensitive means of comparison of microclimates than the more obvious approaches. If the hypothesis of identity between two sites is disproved some indication is available of the nature of the difference. A value of b greater than unity indicates that E_1 has a greater range than E_2, a value less than unity the reverse.

Hocker (1956), in a study of the relation of *Pinus taeda* to climate, has calculated a discriminant function, based on a number of meteorological measures, between stations within the natural limits of the species and those outside the limits but within an arbitrarily defined distance from them. Of the 21 measures considered, only 12 made a significant contribution to the discriminant function. It was thus possible to obtain some information on the aspects of climate that were limiting to the species. While interpretation of such data clearly requires considerable knowledge of the edaphic and other non-climatic tolerances of the species concerned, this approach has evident potentialities in relation to microclimatic as well as macroclimatic data.

2. *Area and spread of species*

Study of the distribution of species within areas relatively narrowly defined climatically, but covering a range of vegetation types, falls on the borderline between ecology and plant geography. Such studies are well exemplified by investigations of the occurrence of species in the 152 vice-counties into which Great Britain has been divided for biological purposes. These vice-counties represent units broadly similar in size and each fairly homogeneous climatically. Generally, comparison of vice-comital distributions can be made on inspection, without recourse to quantitative tests. If two species are suspected of showing similar distribution patterns, a test of association between occurrences may be applied (see Chapter 4) but since similarities of distribution in such large units are generally due to similar climatic preferences, there will generally be blocks of vice-counties occupied or not occupied by the two species and the similarities of distribution will be readily apparent. If it is suspected that a species is more widespread in one part of an area than another, but the difference is not clear on inspection, it is useful to examine the relative occurrences in a contingency table. The hepatic *Fossombronia Wondraczeki* has been reported from 10 out of 41 vice-counties in Scotland, and 23 out of 71 vice-counties in England and Wales. Is there any evidence from these data alone that it is more widespread in England and Wales? The contingency table is:

	F. Wondraczeki		
	present	absent	
Scotland	10	31	41
England and Wales	23	48	71
	33	79	112

This has a corrected χ^2 of 0·761, with probability 30–50 per cent, giving no indication of real difference in occurrence.

In considering regional distribution of species reported from very few vice-counties it is salutary to bear in mind the chance probability of the observed distributions. *Moerkia hibernica* is recorded from two vice-counties (Forfar and Shetland) in Scotland and none in England and Wales, and at first sight it might be regarded as a northern species. However, the contingency table

	M. hibernica		
	present	absent	
Scotland	2	39	41
England and Wales	0	71	71
	2	110	112

has a chance probability of occurrence (by the exact test) of 13·2% and does not support the supposition. In interpreting such data not only the number of vice-counties occupied but also the relative position of those vice-counties that are occupied must be considered. Thus *Marsupella apiculata* is recorded from three vice-counties (South Aberdeen, East Inverness and West Inverness) in Scotland and none in England and Wales. The probability of the contingency table is 4·7% by the exact test, corresponding to a probability by the χ^2 test of 9·4%, and would not normally be considered significant. However, in this case the fact that the three vice-counties of occurrence are adjacent is additional evidence and a hypothesis of northern occurrence is not unreasonable.

223

The cases quoted are put forward as examples rather than for their intrinsic interest. Such distribution problems within the British Isles, or other areas of comparable size, can rarely be interpreted on the internal evidence alone, without taking into consideration their distribution elsewhere. The value of quantitative tests is chiefly as a warning against too ready acceptance of apparent distribution patterns.

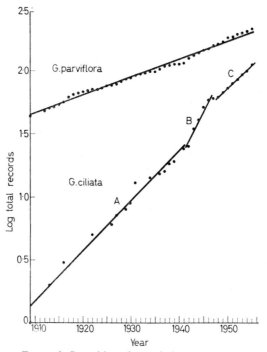

FIGURE 1. Logarithm of cumulative total number of records of *Galinsoga parviflora* and *G. ciliata* in the British Isles from 1909 to 1955. Fitted regression lines are shown. The lines A, B and C for *G. ciliata* are for the three periods 1909–41, 1942–47 and 1948–55 respectively (after Lacey, 1957)

There is another type of distributional problem in which a quantitative approach may be useful, that of the rate of spread of an introduced species. This may be illustrated from Lacey's (1957) data on the spread in the British Isles of the introduced species *Galinsoga parviflora* and *G. ciliata*. The former was first recorded in 1860, the latter in 1909. Lacey assembled data for the cumulative total number of records for each species from 1909 to 1955. The total number of records of a spreading species may be expected to increase exponentially, so that it is appropriate to apply a logarithmic transformation. FIGURE 1 shows the logarithm of the number of records plotted against date. Two questions arise: (1) Have the two species spread at the same rate? (2) Is the apparently greater rate of spread of *G. ciliata* in the years 1942–47 anything more than a chance effect? Inspection of the graphs leaves little doubt that *G. parviflora* has, over the period 1909–55, had a lower rate of spread than *G. ciliata*, and this is confirmed by fitting a regression line to each set of data; the difference in slope corresponds to a probability of chance occurrence of very much less than 0·1 per cent. The difference between the rates of spread of *G. ciliata* in the years 1942–47 and the years before and after can be tested by fitting regression lines for the three periods separately. These are

A 1909–41 $y = 0·1281 + 0·03972x$
B 1942–47 $y = -1·0934 + 0·07686x$
C 1948–55 $y = 0·3832 + 0·03607x$

where y = logarithm of number of records, x = coded date (actual date minus 1909). The three lines are shown in FIGURE 1. The differences in slope between A and B, and between B and C, are both significant (probability 0·1–1 per cent) but that between A and C is not (probability 20–30 per cent). There is thus clear evidence of a greater rate of spread between 1942 and 1947, with the rate falling to its former value again after 1947.

APPENDIX B

Table 1

VARIANCE OF BINOMIAL DISTRIBUTION AFTER ANGULAR TRANSFORMATION
(a) For various values of n
(In part from Robinson, 1955)

n \ p	·5	·4 ·6	·3 ·7	·2 ·8	·1 ·9	Approximate value $\left(\dfrac{820\cdot7}{n} \right)$
2	1012·5	972·0	850·5	648·0	364·5	410·3
3	575·7	566·4	530·8	449·3	288·8	273·6
4	365·6	369·4	372·6	348·8	250·1	205·2
5	253·4	261·4	279·0	282·8	222·5	164·2
9	106·7	110·6	123·3	148·1	148·9	91·2
10	94·0	96·6	106·4	125·9	136·0	82·1
16	55·2	55·7	57·8	65·8	81·6	51·3
20	43·4	43·5	44·5	48·8	63·5	41·0
25	34·3	34·6	35·3	36·7	47·3	32·8
30	28·2	28·4	28·7	29·9	36·7	27·4

(b) Extended table for $n = 5$
(from Greig-Smith, 1961c, by courtesy of *J. Ecol.*)

p	·01	·02	·03	·04	·05	·06	·07	·08	·09	·10
0	33·8	64·5	92·3	117·5	140·2	160·6	178·9	195·1	209·6	222·5
·1	233·5	243·3	251·5	259·1	265·3	270·4	274·7	278·1	280·8	282·8
·2	284·2	285·0	285·4	285·4	284·9	284·2	283·2	282·0	280·5	279·0
·3	277·3	275·5	273·8	271·9	270·0	268·2	266·4	264·7	263·0	261·4
·4	260·0	258·6	257·5	256·4	255·5	254·8	254·1	253·8	253·4	253·4

TABLE 2

MEAN RANGE OF SAMPLES OF DIFFERENT SIZES (after Pearson and Hartley, 1954)
Range expressed in units of population standard deviation

Sample size	Range	Sample size	Range	Sample size	Range	Sample size	Range
2	1·128	9	2·970	16	3·532	23	3·858
3	1·693	10	3·078	17	3·588	24	3·895
4	2·059	11	3·173	18	3·640	25	3·931
5	2·326	12	3·258	19	3·689	50	4·498
6	2·534	13	3·336	20	3·735	100	5·015
7	2·704	14	3·407	21	3·778		
8	2·847	15	3·472	22	3·819		

TABLE 3

$$e^{-m}$$

m	0	0·01	0·02	0·03	0·04	0·05	0·06	0·07	0·08	0·09
0	1·0000	·9900	·9802	·9704	·9608	·9512	·9418	·9324	·9231	·9139
0·1	·9048	·8958	·8869	·8781	·8694	·8607	·8521	·8437	·8353	·8270
0·2	·8187	·8106	·8025	·7945	·7866	·7788	·7710	·7634	·7558	·7483

m	0·0	0·1	0·2	0·3	0·4	0·5	0·6	0·7	0·8	0·9
0	1·0000	·9048	·8187	·7408	·6703	·6065	·5488	·4966	·4493	·4066
1	·3679	·3329	·3012	·2725	·2466	·2231	·2019	·1827	·1653	·1496
2	·1353	·1225	·1108	·1003	·0907	·0821	·0743	·0672	·0608	·0550
3	·0498	·0450	·0408	·0369	·0334	·0302	·0273	·0247	·0224	·0202
4	·0183	·0166	·0150	·0136	·0123	·0111	·0101	·0091	·0082	·0074
5	·0067	·0061	·0055	·0050	·0045	·0041	·0037	·0033	·0030	·0027
6	·0025	·0022	·0020	·0018	·0017	·0015	·0014	·0012	·0011	·0010
7	·00091	·00082	·00075	·00068	·00061	·00055	·00050	·00045	·00041	·00037
8	·00033	·00030	·00027	·00025	·00022	·00020	·00018	·00017	·00015	·00014
9	·00012	·00011	·00010	·00009	·00008	·00007	·00007	·00006	·00005	·00005

Very full tables of the Poisson distribution have been published by the Defense Systems Department, General Electric Company (1962)

TABLE 4

$$\sqrt{\frac{2}{N-1}}$$

N	$\sqrt{\dfrac{2}{N-1}}$	N	$\sqrt{\dfrac{2}{N-1}}$
10	·4714	100	·1421
20	·3640	150	·1159
30	·2626	200	·1003
40	·2265	250	·08962
50	·2020	300	·08179
60	·1841	400	·07080
70	·1703	500	·06331
80	·1591	1000	·04474
90	·1499		

TABLE 5

SIGNIFICANT POINTS FOR $\phi = \dfrac{2n_0\,n_2}{n_1{}^2}$

(n_0, n_1, $n_2 =$ number of quadrats containing 0, 1, 2, individuals respectively) (from Moore, 1953, by courtesy of *Ann. Bot., Lond.*)

$N =$ total number of quadrats

$$R = \frac{n_0 + n_1 + n_2}{N} \times 100$$

The figure given is the mean of ϕ plus twice its standard error, i.e. approximately the 5% point

Mean number per quadrat	0·5	1·0	1·5	2·0	2·5
R	99	92	81	68	54
N 50	2·70	2·40	2·46	2·66	2·98
100	2·16	1·95	1·99	2·13	2·34
200	1·80	1·66	1·68	1·77	1·92
300	1·65	1·53	1·55	1·62	1·74
400	1·55	1·46	1·47	1·54	1·63
500	1·49	1·41	1·42	1·48	1·56

Table 6

Densities Corresponding to Different Percentage Frequencies in Random Distributions

F%	0	1	2	3	4	5	6	7	8	9
0	0·	·0100	·0202	·0304	·0409	·0512	·0608	·0726	·0844	·0943
10	·1053	·1165	·1278	·1392	·1508	·1625	·1744	·1863	·1984	·2108
20	·2231	·2357	·2484	·2614	·2744	·2877	·3011	·3147	·3285	·3425
30	·3567	·3711	·3857	·4005	·4155	·4308	·4463	·4620	·4780	·4943
40	·5108	·5277	·5447	·5621	·5798	·5978	·6161	·6349	·6539	·6733
50	·6931	·7133	·7400	·7560	·7765	·7985	·8210	·8440	·8675	·8916
60	·9163	·9416	·9676	·9942	1·0216	1·0498	1·0788	1·1087	1·1394	1·1712
70	1·2040	1·2379	1·2730	1·3093	1·3471	1·3863	1·4271	1·4697	1·5141	1·5606
80	1·6094	1·6607	1·7148	1·7720	1·8326	1·8971	1·9661	2·0402	2·1203	2·2073
90	2·3026	2·4079	2·5257	2·6593	2·8134	2·9957	3·2189	3·5065	3·9120	4·6052

F%	99·1	99·2	99·3	99·4	99·5	99·6	99·7	99·8	99·9
	4·7105	4·8283	4·9618	5·1160	5·2983	5·5215	5·8091	6·2146	6·9078

229

TABLE 7

95 PER CENT CONFIDENCE LIMITS FOR RANDOM DISTRIBUTION OF THE RATIO OF OBSERVED TO EXPECTED VARIANCE IN ANALYSIS OF PATTERN

(From Greig-Smith, 1961c, by courtesy of *J. Ecol.*)

Degrees of freedom	1	2	3	4	5	6	7	8	9	10	12	14
Lower limit	0·00	0·03	0·07	0·12	0·17	0·21	0·24	0·27	0·30	0·32	0·35	0·40
Upper limit	5·02	3·69	3·12	2·79	2·57	2·41	2·29	2·19	2·11	2·05	1·94	1·87
Degrees of freedom	15	16	18	20	24	28	30	32	36	40	48	56
Lower limit	0·42	0·43	0·45	0·48	0·52	0·55	0·56	0·57	0·59	0·61	0·64	0·66
Upper limit	1·83	1·80	1·75	1·71	1·64	1·59	1·57	1·55	1·51	1·48	1·44	1·40
Degrees of freedom	60	64	72	80	96	112	120	128	144	160	192	224
Lower limit	0·67	0·68	0·70	0·71	0·74	0·75	0·75	0·77	0·78	0·79	0·81	0·82
Upper limit	1·39	1·38	1·35	1·33	1·30	1·27	1·26	1·26	1·24	1·23	1·21	1·19
Degrees of freedom	256	288	320	384	448	480	512	576	640	768	896	960
Lower limit	0·83	0·84	0·85	0·86	0·87	0·88	0·88	0·89	0·89	0·90	0·91	0·91
Upper limit	1·18	1·17	1·16	1·15	1·13	1·13	1·13	1·12	1·11	1·10	1·09	1·09
Degrees of freedom	1024	1152	1280	1536								
Lower limit	0·91	0·92	0·92	0·93								
Upper limit	1·09	1·08	1·08	1·07								

Table 8

PROBABILITY THAT S (FOR τ) ATTAINS OR EXCEEDS A SPECIFIED VALUE. (SHOWN ONLY FOR POSITIVE VALUES. NEGATIVE VALUES OBTAINABLE BY SYMMETRY) (From Kendall, 1948, by courtesy of Charles Griffin & Co. Ltd.)

S	Values of n				S	Values of n		
	4	5	8	9		6	7	10
0	0·625	0·592	0·548	0·540	1	0·500	0·500	0·500
2	0·375	0·408	0·452	0·460	3	0·360	0·386	0·431
4	0·167	0·242	0·360	0·381	5	0·235	0·281	0·364
6	0·042	0·117	0·274	0·306	7	0·136	0·191	0·300
8		0·042	0·199	0·238	9	0·068	0·119	0·242
10		$0\cdot0^283$	0·138	0·179	11	0·028	0·068	0·190
12			0·089	0·130	13	$0\cdot0^283$	0·035	0·146
·14			0·054	0·090	15	$0\cdot0^214$	0·015	0·108
16			0·031	0·060	17		$0\cdot0^254$	0·078
18			0·016	0·038	19		$0\cdot0^214$	0·054
20			$0\cdot0^271$	0·022	21		$0\cdot0^320$	0·036
22			$0\cdot0^228$	0·012	23			0·023
24			$0\cdot0^387$	$0\cdot0^263$	25			0·014
26			$0\cdot0^319$	$0\cdot0^229$	27			$0\cdot0^283$
28			$0\cdot0^425$	$0\cdot0^212$	29			$0\cdot0^246$
30				$0\cdot0^343$	31			$0\cdot0^223$
32				$0\cdot0^312$	33			$0\cdot0^211$
34				$0\cdot0^425$	35			$0\cdot0^347$
36				$0\cdot0^528$	37			$0\cdot0^318$
					39			$0\cdot0^458$
					41			$0\cdot0^415$
					43			$0\cdot0^528$
					45			$0\cdot0^628$

Repeated zeros are indicated by powers, *e.g.* $0\cdot0^347$ stands for $0\cdot00047$

REFERENCES

ABERDEEN, J. E. C. (1954). Estimation of basal or cover areas in plant ecology. *Aust. J. Sci.*, **17**, 35–36.

—— (1955). Quantitative methods for estimating the distribution of soil fungi. *Pap. Dep. Bot. Univ. Qd*, **3**, 83–96.

—— (1958). The effect of quadrat size, plant size and plant distribution on frequency estimates in plant ecology. *Aust. J. Bot.*, **6**, 47–58.

AGNEW, A. D. Q. (1957). The ecology of British rushes with special reference to the invasion of re-seeded pastures. Ph.D. thesis. University of Wales.

—— (1961). The ecology of *Juncus effusus* L. in North Wales. *J. Ecol.*, **49**, 83–102.

AITCHISON, J. and BROWN, J. A. C. (1957). *The lognormal distribution with special reference to its uses in economics.* Cambridge.

ANDERSON, D. J. (1960). A comparison of some upland plant communities with particular reference to their structure. Ph.D. thesis. University of Wales.

—— (1963). The structure of some upland plant communities in Caernarvonshire. III. The continuum analysis. *J. Ecol.*, **51**, 403–414.

ARCHIBALD, E. E. A. (1948). Plant populations. I. A new application of Neyman's contagious distribution. *Ann. Bot., Lond.*, N.S. **12**, 221–235.

—— (1949a). The specific character of plant communities. I. Herbaceous communities. *J. Ecol.*, **37**, 260–273.

—— (1949b). The specific character of plant communities. II. A quantitative approach. *J. Ecol.*, **37**, 274–288.

—— (1950). Plant populations. II. The estimation of the number of individuals per unit area of species in heterogeneous plant populations. *Ann. Bot., Lond.*, N.S. **14**, 7–21.

—— (1952). A possible method for estimating the area covered by the basal parts of plants. *S. Afr. J. Sci.*, **48**, 286–292.

ASHBY, E. (1935). The quantitative analysis of vegetation. *Ann. Bot., Lond.*, **49**, 779–802.

—— (1936). Statistical ecology. *Bot. Rev.*, **2**, 221–235.

—— (1948). Statistical ecology. II. A reassessment. *Bot. Rev.*, **14**, 222–234.

ASHTON, P. S. (1962). Taxonomy and ecology of the Dipterocarpaceae in Brunei. Ph.D. thesis. University of Cambridge.

REFERENCES

BARNES, H. and STANBURY, F. A. (1951). A statistical study of plant distribution during the colonization and early development of vegetation on china clay residues. *J. Ecol.*, **39**, 171–181.

BARTLETT, M. S. (1936). Some examples of statistical methods of research in agriculture and applied biology. *Suppl. J. R. statist. Soc.*, **4**, 137–183.

BEALS, E. (1960). Forest bird communities in the Apostle Islands of Wisconsin. *Wilson Bull.*, **72**, 156–181.

BESCHEL, R. E. and WEBBER, P. J. (1962). Gradient analysis in swamp forests. *Nature, Lond.*, **194**, 207–209.

BIELESKI, R. L. (1959). Factors affecting growth and distribution of Kauri (*Agathis australis* Salisb.). I. Effect of light on the establishment of Kauri and of *Phyllocladus trichomanoides* D. Don. *Aust. J. Bot.*, **7**, 252–267.

BITTERLICK, W. (1948). Die Winkelzählprobe. *Allg. Forst- u. Holzw. Ztg*, **59**, 4–5.

BLACK, G. A., DOBZHANSKY, Th. and PAVAN, C. (1950). Some attempts to estimate species diversity and population density of trees in Amazonian Forests. *Bot. Gaz.*, **111**, 413–425.

BLACKMAN, G. E. (1935). A study by statistical methods of the distribution of species in grassland associations. *Ann. Bot., Lond.*, **49**, 749–777.

—— (1942). Statistical and ecological studies in the distribution of species in plant communities. I. Dispersion as a factor in the study of changes in plant populations. *Ann. Bot., Lond.*, N.S. **6**, 351–370.

BLACKMAN, G. E. and RUTTER, A. J. (1946). Physiological and ecological studies in the analysis of plant environment. I. The light factor and the distribution of the bluebell (*Scilla non-scripta*) in woodland communities. *Ann. Bot., Lond.*, N.S. **10**, 361–390.

BONNER, J. (1950). The rôle of toxic substances in the interactions of higher plants. *Bot. Rev.*, **16**, 51–65.

BORMANN, F. H. (1953). The statistical efficiency of sample plot size and shape in forest ecology. *Ecology*, **34**, 474–487.

BOURDEAU, P. F. (1953). A test of random versus systematic ecological sampling. *Ecology*, **34**, 499–512.

BRAUN-BLANQUET, J. (1951). *Pflanzensoziologie*. 2nd edn. Vienna.

BRAY, J. R. (1956). A study of mutual occurrence of plant species. *Ecology*, **37**, 21–28.

—— (1961). A test for estimating the relative informativeness of vegetation gradients. *J. Ecol.*, **49**, 631–642.

—— (1962). Use of non-area analytic data to determine species dispersion. *Ecology*, **43**, 328–333.

BRAY, J. R. and CURTIS, J. T. (1957). An ordination of the upland forest communities of southern Wisconsin. *Ecol. Monogr.*, **27**, 325–349.

BROWN, D. (1954.) Methods of surveying and measuring vegetation. *Bull. Bur. Past., Hurley*, **42.**

BROWN, R. T. and CURTIS, J. T. (1952). The upland conifer-hardwood forests of northern Wisconsin. *Ecol. Monogr.*, **22**, 217–234.

CAIN, S. A. (1934). Studies on virgin hardwood forest. II. A comparison of quadrat sizes in a quantitative phytosociological study of Nash's Woods, Posey County, Indiana. *Amer. Midl. Nat.*, **15,** 529–566.

—— (1938). The species-area curve. *Amer. Midl. Nat.*, **19,** 573–581.

—— (1943). Sample-plot technique applied to alpine vegetation in Wyoming. *Amer. J. Bot.*, **30,** 240–247.

CATTELL, R. B. (1952). *Factor analysis.* New York.

CHRISTIAN, C. S. and PERRY, R. A. (1953). The systematic description of plant communities by the use of symbols. *J. Ecol.*, **41,** 100–105.

CLAPHAM, A. R. (1932). The form of the observational unit in quantitative ecology. *J. Ecol.*, **20,** 192–197.

—— (1936). Over-dispersion in grassland communities and the use of statistical methods in plant ecology. *J. Ecol.*, **24,** 232–251.

CLARK, P. J. (1956). Grouping in spatial distributions. *Science*, **123,** 373–374.

CLARK, P. J. and EVANS, F. C. (1954a). Distance to nearest neighbour as a measure of spatial relationships in populations. *Ecology*, **35,** 445–453.

—— (1954b). On some aspects of spatial pattern in biological populations. *Science*, **121,** 397–398.

CLAUSEN, J. J. (1957). A comparison of some methods of establishing plant community patterns. *Bot. Tidsskr.*, **53,** 253–278.

CLEMENTS, F. E. (1916). Plant succession: an analysis of the development of vegetation. *Publ. Carneg. Instn.*, **242.**

CLEMENTS, F. E. and SHELFORD, V. E. (1939). *Bio-ecology.* New York and London.

COCHRAN, W. G. (1954). Some methods for strengthening the common χ^2 tests. *Biometrics*, **4,** 417–451.

—— (1963). *Sampling techniques.* 2nd edn. New York and London.

COCHRAN, W. G. and COX, G. M. (1944). *Experimental design.* Mimeographed. (Quoted from Snedecor, 1946.)

COLE, L. C. (1949). The measurement of interspecific association. *Ecology*, **30,** 411–424.

—— (1957). The measurement of partial interspecific association. *Ecology*, **38,** 226–233.

CONWAY, V. M. (1938). Studies in the autecology of *Cladium mariscus* R. Br. IV. Growth rates of the leaves. *New Phytol.*, **37,** 254–278.

COTTAM, G. (1947). A point method for making rapid surveys of woodlands. *Bull. ecol. Soc. Amer.*, **28,** 60.

COTTAM, G. and CURTIS, J. T. (1948). The use of the punched card method in phytosociological research. *Ecology*, **29,** 516–519.

—— (1949). A method for making rapid surveys of woodlands by means of pairs of randomly selected trees. *Ecology*, **30,** 101—104.

—— (1955). Correction for various exclusion angles in the random pairs method. *Ecology*, **36,** 767.

—— (1956). The use of distance measures in phytosociological sampling. *Ecology*, **37,** 451–460.

REFERENCES

COTTAM, G., CURTIS, J. T. and CATANA, A. J. (1957). Some sampling characteristics of a series of aggregated populations. *Ecology*, **38**, 610–622.

COTTAM, G., CURTIS, J. T. and HALE, B. W. (1953). Some sampling characteristics of a population of randomly dispersed individuals. *Ecology*, **34**, 741–757.

CURTIS, J. T. (1947). The palo verde forest type near Gonaives, Haiti, and its relation to the surrounding vegetation. *Caribbean For.*, **8**, 1–26.

—— (1955). A note on recent work dealing with the spatial distribution of plants. *J. Ecol.*, **43**, 309.

—— (1959). *The vegetation of Wisconsin: an ordination of plant communities.* Madison, Wisconsin.

CURTIS, J. T. and GREENE, H. C. (1949). A study of relic Wisconsin prairies by the species-presence method. *Ecology*, **30**, 83–92.

CURTIS, J. T. and McINTOSH, R. P. (1950). The interrelation of certain analytic and synthetic phytosociological characters. *Ecology*, **31**, 434–455.

—— (1951). An upland forest continuum in the prairie-forest border region of Wisconsin. *Ecology*, **32**, 476–496.

CZEKANOWSKI, J. (1913). *Zarys metod statystycznyck.* Warsaw.

DAGNELIE, P. (1960). Contribution à l'étude des communautés végétales par l'analyse factorielle. *Bull. Serv. Carte phytogéogr. Sér.B*, **5**, 7–71 and 93–195.

DAHL, E. (1957). Rondane: mountain vegetation in South Norway and its relation to the environment. *Skr. norske VidenskAkad. I Mat.-Naturv. Klasse*, 1956(3). 1–374.

—— (1960). Some measures of uniformity in vegetation analysis. *Ecology*, **41**, 805–808.

DAVID, F. N. and MOORE, P. G. (1954). Notes on contagious distributions in plant populations. *Ann. Bot., Lond.*, N.S. **18**, 47–53.

—— (1957). A bivariate test for the clumping of supposedly random individuals. *Ann. Bot., Lond.*, N.S. **21**, 315–320.

DAVIS, T. A. W. and RICHARDS, P. W. (1933–4). The vegetation of Moraballi Creek, British Guiana: an ecological study of a limited area of tropical rain forest. *J. Ecol.*, **21**, 350–384, **22**, 106–155.

DAWSON, G. W. P. (1951). A method for investigating the relationship between the distribution of individuals of different species in a plant community. *Ecology*, **32**, 332–334.

DEFENSE SYSTEMS DEPARTMENT GENERAL ELECTRIC COMPANY (1962). *Tables of the individual and cumulative terms of Poisson distribution.* Princeton.

DE VRIES, D. M. (1953). Objective combinations of species. *Acta bot. neerl.*, **1**, 497–499.

—— (1954a). Constellation of frequent herbage plants, based on their correlation in occurrence. *Vegetatio, Haag*, **5–6**, 105–111.

—— (1954b). Ecological results obtained by the use of interspecific correlation. *European Grassland Conference*, pp. 32–36. O.E.E.C., Paris.

De Vries, D. M. and De Boer, T. A. (1959). Methods used in botanical grassland research in the Netherlands and their application. *Herb. Abstr.*, **29**, 1–7.

Dice, L. R. (1945). Measures of the amount of ecologic association between species. *Ecology*, **26**, 297–302.

—— (1952). Measure of the spacing between individuals within a population. *Contr. Lab. Vertebr. Biol. Univ. Mich.*, **55**, 1–23.

Dix, R. L. (1959). The influence of grazing on the thin-soil prairies of Wisconsin. *Ecology*, **40**, 36–49.

—— (1961). An application of the point-centred quarter method to the sampling of grassland vegetation. *J. Range Mgmt.*, **14**, 63–69.

Dix, R. L. and Butler, J. E. (1960). A phytosociological study of a small prairie in Wisconsin. *Ecology*, **41**, 316–327.

Dixon, W. J. and Massey, F. J. (1957). *Introduction to statistical analysis.* 2nd edn. New York.

Du Rietz, G. E. (1921). Zur methodologischen Grundlage de modernen Pflanzensoziologie. *Akad. Afh., Uppsala.*

—— (1931). Life-forms of terrestrial flowering plants. *Acta phytogeogr. suec.*, **3**, 1–95.

Ellenberg, H. (1950–54). *Landwirtschaftliche Pflanzensoziologie.* Stuttgart.

—— (1956). Ausgaben und Methoden der Vegetationskunde. H. Walter, *Einführung in die Phytologie.* IV. Grundlagen der Vegetationsgliederung. I. Teil, 136 pp.

Ellison, L. (1942). A comparison of methods of quadratting short-grass vegetation. *J. agric. Res.*, **64**, 595–614.

Emmett, H. E. G. and Ashby, E. (1934). Some observations on the relation between the hydrogen-ion concentration of the soil and plant distribution. *Ann. Bot., Lond.*, **48**, 869–876.

Erickson, R. O. and Stehn, J. R. (1945). A technique for analysis of population density data. *Amer. midl. Nat.*, **33**, 781–787.

Evans, D. A. (1953). Experimental evidence concerning contagious distributions in ecology. *Biometrika*, **40**, 186–211.

Evans, F. C. (1952). The influence of size of quadrat on the distributional patterns of plant populations. *Contr. Lab. Vertebr. Biol. Univ. Mich.*, **54**, 1–15.

Ewer, S. J. (1932). Life forms of Illinois plants. *Trans. Ill. Acad. Sci.*, **24**, 108–121.

Fager, E. W. (1957). Determination and analysis of recurrent groups. *Ecology*, **38**, 586–595.

Falinski, J. (1960). Zastosowanie taksonomii wroclawskiej do fitosocjologii. (Anwendung der sog. 'Breslauer Taxonomie' in der Pflanzensoziologie.) *Acta Soc. Bot. Polon.*, **29**, 333–361.

Feller, W. (1943). On a general class of 'contagious' distributions. *Ann. math. Statist.*, **14**, 389–400.

Ferrari, H. P., Pijl, H. and Venekamp, J. T. N. (1957). Factor analysis in agricultural research. *Netherlands J. agric. Sci.*, **5**, 211–221.

REFERENCES

FINNEY, D. J. (1948). Random and systematic sampling in timber surveys. *Forestry*, **22**, 64–99.

—— (1950). An example of periodic variation in forest sampling. *Forestry*, **23**, 96–111.

FISHER, R. A. (1936). The use of multiple measurements in taxonomic problems. *Ann. Eugen., Lond.*, **7**, 179–188.

—— (1941). *Statistical methods for research workers*. 8th edn. Edinburgh and London.

FISHER, R. A. and YATES, F. (1943). *Statistical tables for biological, agricultural and medical research*. 2nd edn. London and Edinburgh.

FRACKER, S. B. and BRISCHLE, H. A. (1944). Measuring the local distribution of *Ribes. Ecology*, **25**, 283–303.

FREEMAN, M. F. and TUKEY, J. W. (1950). Transformations related to the angular and the square root. *Ann. math. Statist.*, **21**, 607–611.

FRITTS, H. C. (1960). Multiple regression analysis of radial growth in individual trees. *For. Sci.*, **6**, 334–349.

GAMS, H. (1961). Erfassung und Darstellung mehrdimensionaler Verwandtschaftsbeziehungen von Sippen und Lebensgemeinschaften. *Ber. Geobot. Inst. ETH, Stiftg. Rubel, Zürich*, **32**, 96–116.

GHENT, A. W. (1963). Kendall's 'tau' coefficient as an index of similarity in comparisons of plant or animal communities. *Canad. Ent.* **95**, 568–575.

GILBERT, M. L. and CURTIS, J. T. (1953). Relation of the understory to the upland forest in the prairie-forest border region of Wisconsin. *Trans. Wis. Acad. Sci. Arts Lett.*, **42**, 183–195.

GIMINGHAM, C. H. (1961). Northern European heath communities: a 'network of variation'. *J. Ecol.*, **49**, 655–694.

GIMINGHAM, C. H., GEMMELL, A. R. and GREIG-SMITH, P. (1948). The vegetation of a sand-dune system in the Outer Hebrides. *Trans. bot. Soc. Edinb.*, **25**, 82–96.

GITTINS, R. T. (1963). A study of the structure of some grassland and heath communities in Anglesey. Ph.D. thesis. University of Wales.

GLEASON, H. A. (1920). Some applications of the quadrat method. *Bull. Torrey bot. Cl.*, **47**, 21–33.

—— (1922). On the relation between species and area. *Ecology*, **3**, 158–162.

—— (1925). Species and area. *Ecology*, **6**, 66–74.

—— (1926). The individualistic concept of the plant association. *Bull. Torrey bot. Cl.*, **53**, 7–26.

GOODALL, D. W. (1952a). Quantitative aspects of plant distribution. *Biol. Rev.*, **27**, 194–245.

—— (1952b). Some considerations in the use of point quadrats for the analysis of vegetation. *Aust. J. sci. Res. Ser. B*, **5**, 1–41.

—— (1953a). Objective methods for the classification of vegetation. I. The use of positive interspecific correlation. *Aust. J. Bot.*, **1**, 39–63.

—— (1953b). Objective methods for the classification of vegetation. II. Fidelity and indicator value. *Aust. J. Bot.*, **1**, 434–456.

—— (1953c). Point quadrat methods for the analysis of vegetation. The treatment of data for tussock grasses. *Aust. J. Bot.*, **1**, 457–461.

GOODALL, D. W. (1954a). Vegetational classification and vegetational continua. *Angew. PflanzSoz.*, **1**, 168–182.

—— (1954b). Minimal area: a new approach. *Int. bot. Congr. 8, Rap.* Sect. 7 and 8, 19–21.

—— (1954c). Objective methods for the classification of vegetation. III. An essay in the use of factor analysis. *Aust. J. Bot.*, **2**, 304–324.

—— (1961). Objective methods for the classification of vegetation. IV. Pattern and minimal area. *Aust. J. Bot.*, **9**, 162–196.

—— (1962). Bibliography of statistical plant sociology. *Excerpta bot.* Sect. B, **4**, 253–322.

GREIG-SMITH, P. (1952a). The use of random and contiguous quadrats in the study of the structure of plant communities. *Ann. Bot., Lond.*, N.S. **16**, 293–316.

—— (1952b). Ecological observations on degraded and secondary forest in Trinidad, British West Indies. II. Structure of the communities. *J. Ecol.*, **40**, 316–330.

—— (1961a) The use of pattern analysis in ecological investigations. *Recent advances in botany*, **2**, pp. 1354–1358. Toronto.

—— (1961b). Review of *The Vegetation of Wisconsin* by J. T. Curtis. *J. Ecol.*, **49**, 463–465.

—— (1961c). Data on pattern within plant communities. I. The analysis of pattern. *J. Ecol.*, **49**, 695–702.

—— (1961d). Data on pattern within plant communities. II. *Ammophila arenaria* (L.) Link. *J. Ecol.*, **49**, 703–708.

GREIG-SMITH, P., KERSHAW, K. A. and ANDERSON, D. J. (1963). The analysis of pattern in vegetation: a comment on a paper by D. W. Goodall. *J. Ecol.*, **51**, 223–229.

GROSENBAUGH, L. R. (1952a). Plotless timber estimates—new, fast, easy. *J. For.*, **50**, 32–37.

—— (1952b). Shortcuts for cruisers and scalers. *Occ. Pap. Sth. For. Exp. Sta.*, **126**, 1–24.

GUINOCHET, M. and CASAL, P. (1957). Sur l'analyse différentielle de Czekanowski et son application à la phytosociologie. *Bull. Serv. Carte phytogéogr. Sér.B*, **2**, 25–33.

HAMILTON, K. C. and BUCHHOLTZ, K. P. (1955). Effect of rhizomes of quackgrass (*Agropyron repens*) and shading on the seedling development of weedy species. *Ecology*, **36**, 304–308.

HANSON, H. C. (1934). A comparison of methods of botanical analysis of the native prairie in western North Dakota. *J. agric. Res.*, **49**, 815–842.

HARBERD, D. J. (1960). Association analysis in plant communities. *Nature, Lond.*, **185**, 53–54.

—— (1962). Application of a multivariate technique to ecological survey. *J. Ecol.*, **50**, 1–17.

HARRIS, J. A., KUENZEL, J. and COOPER, W. S. (1929). Comparison of the physical factors of habitats. *Ecology*, **10**, 47–66.

HASEL, A. A. (1938). Sampling error in timber surveys. *J. agric. Res.*, **57**, 713–736.

REFERENCES

HOCKER, H. W. (1956). Certain aspects of climate as related to the distribution of Loblolly pine. *Ecology*, **37**, 824–834.

HOLZINGER, K. J. and HARMAN, H. H. (1941). *Factor Analysis. A synthesis of factorial methods.* University of Chicago Press, Chicago.

HOPE-SIMPSON, J. F. (1940). On the errors in the ordinary use of subjective frequency estimations in grassland. *J. Ecol.*, **28**, 193–209.

HOPKINS, B. (1954). A new method for determining the type of distribution of plant individuals. *Ann. Bot., Lond.*, N.S. **18**, 213–227.

—— (1955). The species-area relations of plant communities. *J. Ecol.*, **43**, 409–426.

—— (1957). Pattern in the plant community. *J. Ecol.*, **45**, 451–463.

HORA, F. B. (1947). The pH range of some cliff plants on rocks of different geological origin in the Cader Idris area of North Wales. *J. Ecol.*, **35**, 158–165.

HUGHES, R. E. (1961). The application of certain aspects of multivariate analysis to plant ecology. *Recent Advances in Botany*, **2**, pp. 1350–1354. Toronto.

HUGHES, R. E. and LINDLEY, D. V. (1955). Application of biometric methods to problems of classification in ecology. *Nature, Lond.*, **175**, 806–807.

HUSTICH, I. (1960). Symposium on forest types and forest ecosystems. *Silva fenn.*, **105**, 1–142.

HUTCHINGS, S. S. and HOLMGREN, R. C. (1959). Interpretation of loop-frequency data as a measure of plant cover. *Ecology*, **40**, 668–677.

JACCARD, P. (1912). The distribution of the flora in the alpine zone. *New Phytol.*, **11**, 37–50.

JOHNSTON, A. (1957). A comparison of the line interception, vertical point quadrat, and loop methods as used in measuring basal area of grassland vegetation. *Canad. J. plant Sci.*, **37**, 34–42.

JONES, E. W. (1945). The regeneration of Douglas Fir, *Pseudotsuga taxifolia* Britt., in the New Forest. *J. Ecol.*, **33**, 44–56.

—— (1955–6). Ecological studies on the rain forest of southern Nigeria. IV. The plateau forest of the Okomu Forest Reserve. *J. Ecol.*, **43**, 564–594, **44**, 83–117.

JOWETT, G. H. and SCURFIELD, G. (1949a). Statistical test for optimal conditions: note on a paper of Emmett and Ashby. *J. Ecol.*, **37**, 65–67.

—— (1949b). A statistical investigation into the distribution of *Holcus mollis* L. and *Deschampsia flexuosa* (L.) Trin. *J. Ecol.*, **37**, 68–81.

KEMP, C. D. and KEMP, A. W. (1956). The analysis of point quadrat data. *Aust. J. Bot.*, **4**, 167–174.

KENDALL, M. G. (1948). *Rank correlation methods.* London.

—— (1957). *A course in multivariate analysis.* London.

KERSHAW, K. A. (1957a). The use of cover and frequency in the detection of pattern in plant communities. *Ecology*, **38**, 291–299.

—— (1957b). A study of pattern in certain plant communities. Ph.D. thesis. University of Wales.

—— (1958). An investigation of the structure of a grassland community. I. The pattern of *Agrostis tenuis*. *J. Ecol.*, **46**, 571–592.

KERSHAW, K. A. (1959). An investigation of the structure of a grassland community. II. The pattern of *Dactylis glomerata, Lolium perenne* and *Trifolium repens*. III. Discussion and conclusions. *J. Ecol.*, **47,** 31–53.

—— (1960). The detection of pattern and association. *J. Ecol.*, **48,** 233–242.

—— (1961). Association and co-variance analysis of plant communities. *J. Ecol.*, **49,** 643–654.

—— (1962a). Quantitative ecological studies from Landmannahellir, Iceland. I. *Eriophorum angustifolium. J. Ecol.*, **50,** 163–169.

—— (1962b). Quantitative ecological studies from Landmannahellir, Iceland. III. Variation of performance in *Carex bigelowii. J. Ecol.*, **50,** 393–399.

KING, J. (1962). The *Festuca–Agrostis* grassland complex in south-east Scotland. *J. Ecol.*, **50,** 321–356.

KITTREDGE, J. (1938). The interrelations of habitat, growth rate, and associated vegetation in the aspen communities of Minnesota and Wisconsin. *Ecol. Mongr.*, **8,** 151–246.

KOCH, L. F. (1957). Index of biotal dispersity. *Ecology*, **38,** 145–148.

KRISHNA IYER, P. V. (1948). The theory of probability distribution of points on a line. *J. Indian Soc. agric. Statist.*, **1,** 173–195.

—— (1950). The theory of probability distributions of points on a lattice. *Ann. math. Statist.*, **21,** 198–217.

KYLIN, H. (1926). Über Begriffsbildung und Statistik in der Pflanzensoziologie. *Bot. Notiser* (1926), 81–180.

LACEY, W. S. (1957). A comparison of the spread of *Galinsoga parviflora* and *G. ciliata* in Britain. In *Progress in the study of the British flora*, Ed. J. E. Lousley, pp. 109–115 (*B.S.B.I. Conference Reports*, **5**). Arbroath.

LAMBERT, J. M. and WILLIAMS, W. T. (1962). Multivariate methods in plant ecology. IV. Nodal analysis. *J. Ecol.*, **50,** 775–802.

LEVY, E. B. (1933). Technique employed in grassland research in New Zealand. 1. Strain testing and strain building. *Bull. Bur. Pl. Genet. Aberystw.*, **11,** 6–16.

LEVY, E. B. and MADDEN, E. A. (1933). The point method of pasture analysis. *N.Z.J. Agric.*, **46,** 267–279.

LINDSEY, A. A., BARTON, J. D. and MILES, S. R. (1958). Field efficiencies of forest sampling methods. *Ecology*, **39,** 428–444.

LOOMAN, J. and CAMPBELL, J. B. (1960). Adaptation of Sørensen's K (1948) for estimating unit affinities in prairie vegetation. *Ecology*, **41,** 409–416.

LOUCKS, O. L. (1962). Ordinating forest communities by means of environmental scalars and phytosociological indices. *Ecol. Monogr.*, **32,** 137–166.

LYNCH, D. W. and SCHUMACHER, F. X. (1941). Concerning the dispersion of natural regeneration. *J. For.*, **39,** 49–51.

McGINNIES, W. G. (1934). The relation between frequency index and abundance as applied to plant populations in a semi-arid region. *Ecology*, **15,** 263–282.

REFERENCES

McIntosh, R. P. (1962). Pattern in a forest community. *Ecology*, **43,** 25–33.

McIntyre, G. A. (1953). Estimation of plant density using line transects. *J. Ecol.*, **41,** 319–330.

McVean, D. N. and Ratcliffe, D. A. (1962). *Plant communities of the Scottish Highlands.* London.

Maillefer, A. (1929). Le coefficient générique de P. Jaccard et sa signification. *Mém. Soc. vaud. Sci. nat.*, **3,** 113–183.

Mainland, D., Herrera, L., and Sutcliffe, M. I. (1956). *Statistical tables for use with binomial samples—contingency tests, confidence limits and sample size estimates.* New York University College of Medicine.

Mather, K. (1946). *Statistical analysis in biology.* 2nd edn. London.

Matuszkiewicz, W. (1948). Roslinnosc lasow okolic lwowa (The vegetation of the forests of the environs of Lvov). *Ann. Univ. M. Curie-Sklodowska*, Sect. C, **3,** 119–193.

Maycock, P. F. and Curtis, J. T. (1960). The phytosociology of boreal conifer-hardwood forests of the Great Lakes region. *Ecol. Monogr.*, **30,** 1–35.

Medin, D. E. (1960). Physical site factors influencing annual production of true mountain mahogany, *Cercocarpus montanus. Ecology*, **41,** 454–460.

Moore, P. G. (1953). A test for non-randomness in plant populations. *Ann. Bot., Lond.*, N.S. **17,** 57–62.

——— (1954). Spacing in plant populations. *Ecology*, **35,** 222–227.

Morisita, M. (1954). Estimation of population density by spacing method. *Mem. Fac. Sci. Kyushu Univ.* Ser. E., **1,** 187–197.

——— (1957). A new method for the estimation of density by the spacing method applicable to non-randomly distributed populations. *Seiro-Seitai*, **7,** 134–144.

——— (1959a). Measuring of the dispersion of individuals and analysis of the distributional patterns. *Mem. Fac. Sci. Kyushu Univ.* Ser. E, **2,** 215–235.

——— (1959b). Measuring of interspecific association and similarity between communities. *Mem. Fac. Sci. Kyushu Univ.* Ser. E, **3,** 65–80.

Moroney, M. J. (1951). *Facts from figures.* Penguin Books, Harmondsworth.

Mosteller, F. and Youtz, C. (1961). Tables of the Freeman-Tukey transformations for the binomial and Poisson distributions. *Biometrika*, **48,** 433–440.

Motomura, I. (1952). Comparison of communities based on correlation coefficients. *Ecol. Rev.*, **13,** 67–71. (Quoted from Dagnelie, 1960.)

Mountford, M. D. (1961). On E. C. Pielou's index of non-randomness. *J. Ecol.*, **49,** 271–276.

Muller, C. H. (1953). The association of desert annuals with shrubs. *Amer. J. Bot.*, **40,** 53–60.

Myers, E. and Chapman, V. J. (1953). Statistical analysis applied to a vegetation type in New Zealand. *Ecology*, **34,** 175–185.

Nash, C. B. (1950). Associations between fish species in tributaries and shore waters of western Lake Erie. *Ecology*, **31,** 561–566.

241

NEYMAN, J. (1939). On a new class of contagious distributions applicable in entomology and bacteriology. *Ann. math. Statist.*, **10**, 35–57.

NIELEN, G. C. J. F. and DIRVEN, J. G. P. (1950). De nauwkeurigheid van de plantensociologische 1/4 dm² frequentie-methode. (The accuracy of the 25 cm² specific frequency method). *Versl. landbouwk. Onderz*, **56** (13), 1–27.

NUMATA, M. (1949). The basis of sampling in the statistics of plant communities.—Studies on the structure of plant communities. III. *Bot. Mag., Tokyo*, **62**, 35–38.

—— (1954). Some special aspects of the structural analysis of plant communities. *J. Coll. Arts Sci., Chiba Univ.*, **1**, 194–202.

OGAWA, H., YODA, K. and KIRA, T. (1961). A preliminary survey on the vegetation of Thailand. In *Nature and life in southeast Asia* ed. T. Kira and T. Umesao, vol. **I**, pp. 22–157. Kyoto.

OSVALD, H. (1947). Växternas vapen i kampen om utrymet. *Växtodling*, **2**, 288–303.

PALLEY, M. N. and HORWITZ, L. G. (1961). Properties of some random and systematic point sampling estimators. *For. Sci.*, **7**, 52–65.

PARKER, K. W. (1950). Report on 3-step method for measuring condition and trend of forest ranges. U.S. Forest Service, Washington D.C., 68pp. (processed). (Quoted from Hutchings and Holmgren, 1959.)

—— (1951). A method for measuring trend in range condition on national forest ranges. U.S. Forest Service, Washington D.C., 26pp. (processed). (Quoted from Hutchings and Holmgren, 1959.)

PEARCE, S. C. (1958). Some recent applications of multivariate analysis to data from fruit trees. *Rep. E. Malling Res. Sta.*, 1958, 73–76.

PEARSALL, W. H. (1924). The statistical analysis of vegetation: a criticism of the concepts and methods of the Uppsala school. *J. Ecol.*, **12**, 135–139.

PEARSON, E. S. and HARTLEY, H. O. (1954). *Biometrika tables for statisticians.* Vol. 1. 4th edn. Cambridge.

PEARSON, K. (1934). *Tables of the Incomplete Beta-Function.* Biometrika, London.

PECHANEC, J. F. and STEWART, G. (1940). Sagebrush-grass range sampling studies: size and structure of sampling unit. *J. Amer. Soc. Agron.*, **32**, 669–682.

PHILLIPS, J. (1934–5). Succession, development, the climax and the complex organism: an analysis of concepts. *J. Ecol.*, **22**, 554–571, **23**, 210–246, 488–508.

PHILLIPS, M. E. (1953). Studies in the quantitative morphology and ecology of *Eriophorum angustifolium* Roth. I. The rhizome system. *J. Ecol.*, **41**, 295–318.

—— (1954a). Studies in the quantitative morphology and ecology of *Eriophorum angustifolium* Roth. II. Competition and dispersion. *J. Ecol.*, **42**, 187–210.

—— (1954b). Studies in the quantitative morphology and ecology of *Eriophorum angustifolium* Roth. III. The leafy shoot. *New Phytol.*, **53**, 312–343.

REFERENCES

PIDGEON, I. M. and ASHBY, E. (1940). Studies in applied ecology. I. A statistical analysis of regeneration following protection from grazing. *Proc. Linn. Soc. N.S.W.*, **65**, 123–143.

—— (1942). A new quantitative method of analysis of plant communities. *Aust. J. Sci.*, **5**, 19–21.

PIELOU, E. C. (1957). The effect of quadrat size on the estimation of the parameter of Neyman's and Thomas's distribution. *J. Ecol.*, **45**, 31–47.

—— (1959). The use of point-to-plant distances in the study of the pattern of plant populations. *J. Ecol.*, **47**, 607–613.

—— (1960). A single mechanism to account for regular, random and aggregated populations. *J. Ecol.*, **48**, 575–584.

—— (1961). Segregation and symmetry in two-species populations as studied by nearest neighbour relations. *J. Ecol.*, **49**, 255–269.

—— (1962a). The use of plant-to-neighbour distances for the detection of competition. *J. Ecol.*, **50**, 357–368.

—— (1962b). Runs of one species with respect to another in transects through plant populations. *Biometrics*, **18**, 579–593.

PÓLYA, G. (1930). Sur quelques points de la théorie des probabilités. *Ann. Inst. Poincaré*, **1**, 117–161.

POORE, M. E. D. (1955a). The use of phytosociological methods in ecological investigations. I. The Braun-Blanquet system. *J. Ecol.*, **43**, 226–244.

—— (1955b). The use of phytosociological methods in ecological investigations. II. Practical issues involved in an attempt to apply the Braun-Blanquet system. *J. Ecol.*, **43**, 245–269.

—— (1955c). The use of phytosociological methods in ecological investigations. III. Practical applications. *J. Ecol.*, **43**, 606–651.

—— (1956). The use of phytosociological methods in ecological investigations. IV. General discussion of phytosociological problems. *J. Ecol.*, **44**, 28–50.

PRESTON, F. W. (1948). The commonness, and rarity, of species. *Ecology*, **29**, 254–283.

—— (1960). Time and space and the variation of species. *Ecology*, **41**, 611–627.

—— (1962). The canonical distribution of commonness and rarity. *Ecology*, **43**, 185–215 and 410–431.

RAABE, E. W. (1952). Uber den 'Affinitatswert' in der Planzensoziologie. *Vegetatio, Haag*, **4**, 53–68.

RAO, C. R. (1952). *Advanced statistical methods in biometric research.* New York and London.

RAUNKIAER, C. (1934). *The life forms of plants and statistical plant geography.* Oxford.

RICE, E. L. and PENFOUND, W. T. (1955). An evaluation of the variable-radius and paired-tree methods in the black-jack post oak forest. *Ecology*, **36**, 315–320.

RICHARDS, P. W. (1952). *The tropical rain forest.* Cambridge.

243

RICHARDS, P. W., TANSLEY, A. G. and WATT, A. S. (1940). The recording of structure, life form and flora of tropical forest communities as a basis for their classification. *J. Ecol.*, **28**, 224–239.

ROBINSON, P. (1954). The distribution of plant populations. *Ann. Bot., Lond.,* N.S. **18**, 35–45.

—— (1955). The estimation of ground cover by the point quadrat method. *Ann. Bot., Lond.,* N.S. **19**, 59–66.

RUTTER, A. J. (1955). The composition of wet-heath vegetation in relation to the water-table. *J. Ecol.,* **43**, 507–543.

SALISBURY, E. J. (1925). The incidence of species in relation to soil reaction. *J. Ecol.,* **13**, 149–160.

—— (1939). Ecological aspects of meteorology. *Quart. J. R. met. Soc.,* **65**, 337–358.

SAMPFORD, M. R. (1962.) *An introduction to sampling theory with applications to agriculture.* Edinburgh.

SHANKS, R. E. (1954). Plotless sampling trials in Appalachian forest types. *Ecology,* **35**, 237–244.

SIMPSON, E. H. (1949). Measurement of diversity. *Nature, Lond.,* **163**, 688.

SINGH, B. N. and DAS, K. (1938). Distribution of weed species on arable land. *J. Ecol.,* **26**, 455–466.

SKELLAM, J. G. (1952). Studies in statistical ecology. I. Spatial pattern. *Biometrika,* **39**, 346–362.

SMITH, A. D. (1944). A study of the reliability of range vegetation estimates. *Ecology,* **25**, 441–448.

SNEATH, P. H. A. and SOKAL, R. R. (1962). Numerical taxonomy. *Nature, Lond.,* **193**, 855–860.

SNEDECOR, G. W. (1946). *Statistical methods applied to experiments in agriculture and biology.* 4th edn. Ames, Iowa.

SØRENSEN, T. (1948). A method of establishing groups of equal amplitude in plant sociology based on similarity of species content. *K. danske vidensk. Selsk.,* **5** (4), 1–34.

STEIGER, T. L. (1930). Structure of prairie vegetation. *Ecology,* **11**, 170–217.

STEVENS, W. L. (1937). Significance of grouping. *Ann. Eug., Lond.,* **8**, 57–69.

STEWART, G. and KELLER, W. (1936). A correlation method for ecology as exemplified by studies of native desert vegetation. *Ecology,* **17**, 500–514.

'STUDENT' (1919). An explanation of deviations from Poisson's law in practice. *Biometrika,* **12**, 211–215.

SVEDBERG, T. (1922). Ett bidrag till de statistika metodernas användning inom växtbiologien. *Svensk bot. Tidskr.,* **16**, 1–8.

TANSLEY, A. G. (1920). The classification of vegetation and the concept of development. *J. Ecol.,* **8**, 118–149.

THOMAS, M. (1949). A generalization of Poisson's binomial limit for use in ecology. *Biometrika,* **36**, 18–25.

THOMPSON, H. R. (1955). Spatial point processes, with applications to ecology. *Biometrika,* **42**, 102–115.

REFERENCES

THOMPSON, H. R. (1956). Distribution of distance to *n*th neighbour in a population of randomly distributed individuals. *Ecology*, **37**, 391–394.

—— (1958). The statistical study of plant distribution patterns using a grid of quadrats. *Aust. J. Bot.*, **6**, 322–343.

THOMSON, G. W. (1952). Measures of plant aggregation based on contagious distributions. *Contr. Lab. Vertebr. Biol. Univ. Mich.*, **53**, 1–16.

THURSTONE, L. L. (1947). *Multiple-factor analysis.* Chicago.

TIDMARSH, C. E. M. and HAVANGA, C. M. (1955). The wheel-point method of survey and measurement of semi-open grasslands and karoo vegetation in South Africa. *Mem. bot. Surv. S. Afr.*, **29**, pp. iv+49.

TINNEY, F. W., AAMODT, O. S. and AHLGREN, H. L. (1937). Preliminary report of a study on methods used in botanical analyses of pasture swards. *J. Amer. Soc. Agron.*, **29**, 835–840.

TORGERSON, W. S. (1958). *Theory and method of scaling.* New York.

TRESNER, H. D., BACKUS, M. P. and CURTIS, J. T. (1954). Soil microfungi in relation to the hardwood forest continuum in southern Wisconsin. *Mycologia*, **46**, 314–333.

TUOMIKOSKI, R. (1942). Untersuchungen über die Vegetation der Bruch-moore in Ostfinnland. I. Zur Methodik der pflanzensoziologischen Systematik. *Ann. bot. Soc. Zool.-bot. fenn. Vanamo*, **17** (1), 1–203.

VASILEVICH, V. I. (1961). Association between species and the structure of a phytocoenosis. *Doklady Biol. Sec.*, **139**, 1001–1004.

VESTAL, A. G. (1949). Minimum areas for different vegetations. *Illinois biol. Monogr.*, **20** (3).

VOLK, O. H. (1931). Beiträge zur Ökologie der Sandvegetation de oberrheinischen Tiefebene. *Z. Bot.*, **24**, 81–185.

WATT, A. S. (1940–56). Contributions to the ecology of bracken (*Pteridium aquilinum*). *New Phytol.*, **39**, 401–422; **42**, 103–126; **44**, 156–178; **46**, 98–121; **49**, 308–327; **53**, 117–130; **55**, 369–381.

—— (1947). Pattern and process in the plant community. *J. Ecol.*, **35**, 1–22.

—— (1955). Bracken versus heather, a study in plant sociology. *J. Ecol.*, **43**, 490–506.

WARREN WILSON, J. (1959a). Analysis of the spatial distribution of foliage by two-dimensional point quadrats. *New Phytol.*, **58**, 92–101.

—— (1959b). Analysis of the distribution of foliage area in grassland. In *The measurement of grassland productivity*, ed. J. D. Ivins, pp. 51–61. London.

—— (1960). Inclined point quadrats. *New Phytol.*, **59**, 1–8.

WEBB, D. A. (1954). Is the classification of plant communities either possible or desirable? *Bot. Tidsskr.*, **51**, 362–370.

WELCH, J. R. (1960). Observations on deciduous woodland in the Eastern Province of Tanganyika. *J. Ecol.*, **48**, 557–573.

WENT, F. W. (1942). The dependence of certain annual plants on shrubs in Southern California deserts. *Bull. Torrey. bot. Cl.*, **69**, 100–114.

WEST, O. (1937). An investigation of the methods of botanical analysis of pasture. *S. Afr. J. Sci.*, **33**, 501–559.

245

WHITFORD, P. B. (1949). Distribution of woodland plants in relation to succession and clonal growth. *Ecology*, **30,** 199–208.

WHITTAKER, R. H. (1952). A study of summer foliage insect communities in the Great Smoky Mountains. *Ecol. Monogr.*, **22,** 1–44.

—— (1953). A consideration of climax theory: the climax as a population and pattern. *Ecol. Monogr.*, **23,** 41–78.

—— (1954). Plant populations and the basis of plant indication. *Angew. Pflanzensoz. (Wien)*, *Festschr. Aichinger*, **1,** 183–206.

—— (1956). Vegetation of the Great Smoky Mountains. *Ecol. Monogr.*, **26,** 1–80.

—— (1960). Vegetation of the Siskiyou Mountains, Oregon and California. *Ecol. Monogr.*, **30,** 279–338.

WILLIAMS, C. B. (1947). The logarithmic series and the comparison of island floras. *Proc. Linn. Soc. Lond.*, **158,** 104–108.

—— (1949). Jaccard's generic coefficient and coefficient of floral community, in relation to the logarithmic series and the index of diversity. *Ann. Bot., Lond.*, N.S. **13,** 53–58.

—— (1950). The application of the logarithmic series to the frequency of occurrence of plant species in quadrats. *J. Ecol.*, **38,** 107–138.

WILLIAMS, W. T. and LAMBERT, J. M. (1959). Multivariate methods in plant ecology. I. Association-analysis in plant communities. *J. Ecol.*, **47,** 83–101.

—— (1960). Multivariate methods in plant ecology. II. The use of an electronic digital computer for association-analysis. *J. Ecol.*, **48,** 689–710.

—— (1961a). Multivariate methods in plant ecology. III. Inverse association-analysis. *J. Ecol.*, **49,** 717–729.

—— (1961b). Nodal analysis of associated populations. *Nature, Lond.*, **191,** 202.

—— (1961c). Multivariate methods in taxonomy. (Report of symposium.) *Taxon*, **10,** 205–211.

WILLIAMS, W. T. and LANCE, G. N. (1958). Automatic subdivision of associated populations. *Nature, Lond*,. **182,** 1755.

WINKWORTH, R. E. (1955). The use of point quadrats for the analysis of heathland. *Aust. J. Bot.*, **3,** 68–81.

WINKWORTH, R. E. and GOODALL, D. W. (1962). A crosswise sighting tube for point quadrat analysis. *Ecology*, **43,** 342–343.

INDICES

AUTHOR INDEX

AUTHOR INDEX

SUBJECT INDEX